DATE DUE

DEMCO, INC. 38-2931

Bob Vila's
This Old House

Bob Vila's This Old House

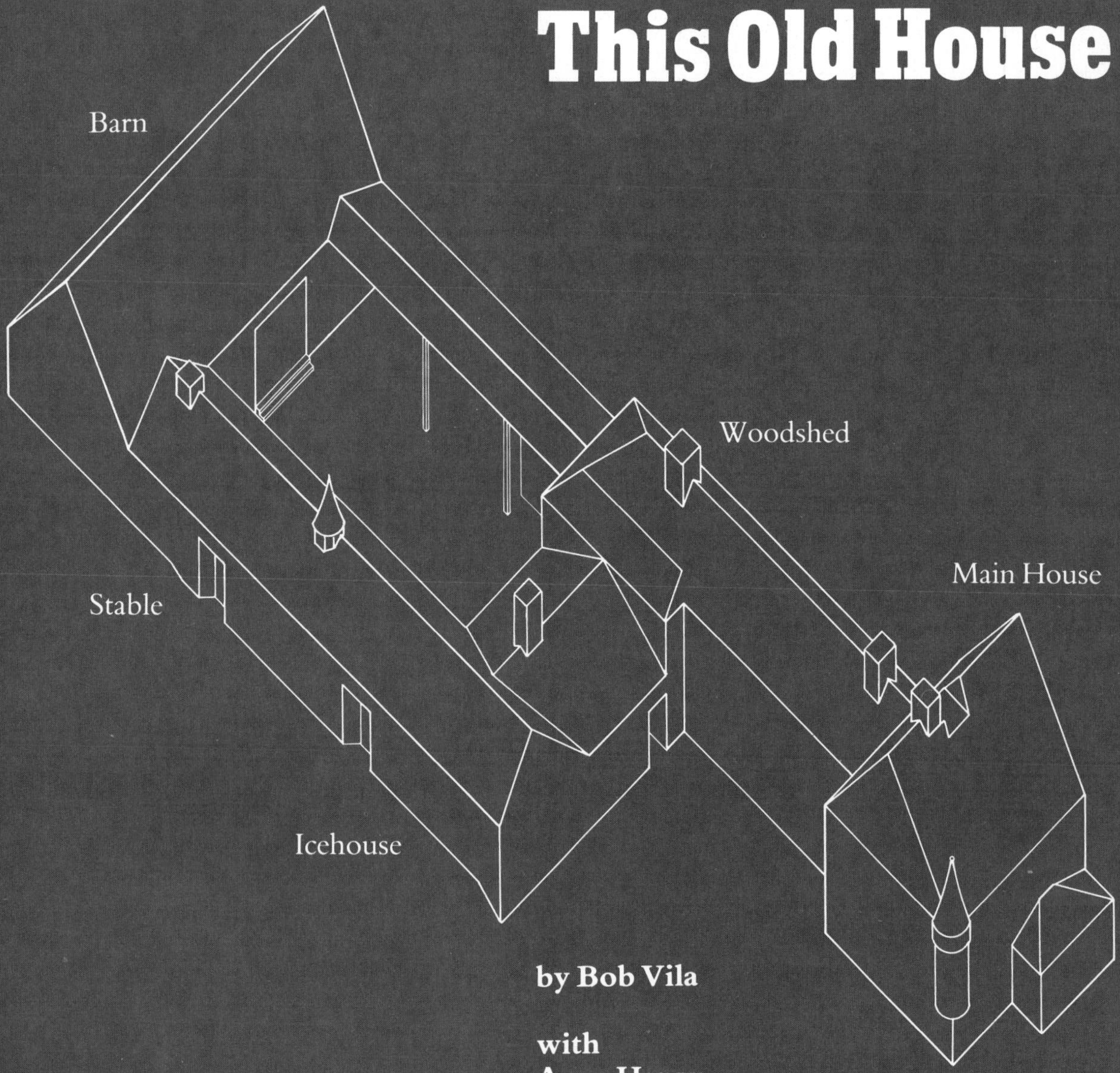

Barn

Woodshed

Main House

Stable

Icehouse

by Bob Vila

with
Anne Henry
Jay Howland
George Stephen
Susan Winston Leff

Photographs by
Bill Schwob

Illustrations by
Anne Henry

E. P. Dutton
New York

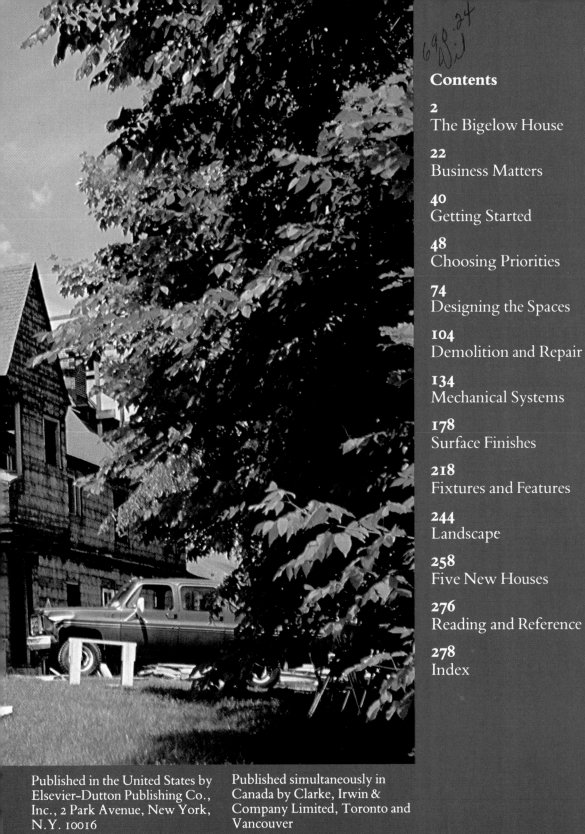

Contents

Published in the United States by Elsevier-Dutton Publishing Co., Inc., 2 Park Avenue, New York, N.Y. 10016

Published simultaneously in Canada by Clarke, Irwin & Company Limited, Toronto and Vancouver

Library of Congress Catalog Card Number: 81–65115

10 9 8 7 6 5 4 3 2 1

First Edition

ISBN: 0–525–93192–0 (cl)
        0–525–47670–9 (pa)

Chapter I

The Bigelow House

by Russell Morash
Producer/Director

World War I truly put an end to the Victorian era. It was the first mass war in history, and it seems that afterward, mass production and mass consumption defined, more and more, the way people lived their lives.

"A civilization is as fragile as a life," Frenchman Paul Éluard wrote. Today, the monuments of the Victorian age are quickly passing away. And the more we see of most modern housing, the more we value the quality of construction and spacious design of an earlier era.

WGBH's first project on "This Old House" was a Victorian single-family house located in the Dorchester section of Boston, a neighborhood that has seen some ups and downs. By the time we had finished the house, the series had attracted a huge audience and everyone involved with the project was happy with the results. Now we had underwriting for a new series lined up (from Montgomery Ward) and as Producer/Director, I found myself in the unenviable position of having to dream up a great new idea, to follow my own act. We finally decided that for the second series we would not repeat the single-family home renovation, but instead, rehabilitate and adapt some other kind of structure. We ended up with another Victorian, but of a very different kind.

2

Wreckage has too often been considered the most economical "conversion" of old buildings, but with costs of new construction going way up, it is no longer always true.

This typical Victorian mansion in Boston served for many years as a private asylum and was scheduled for demolition when the hospital closed. Along with the other hospital buildings, it has been saved by development into condominiums.

The space in this old wharf warehouse has been transformed into a Children's Museum and Museum of Transportation.

All across the country there are run-down shells of buildings—factories, once bustling with machinery and workers, silenced by an age of computers. Large estates, built in the days before a graduated income tax, lie abandoned, lonely ghosts from the nineteenth century.

Once obvious candidates for the wrecker's ball and the bulldozer, these buildings are now being snapped up by people who are interested in adapting them for modern life. Inflation, the baby boom, and changes in attitude and lifestyle have caused people to look for alternatives to the single-family house on its own plot of ground. And more and more they are turning to imaginative re-use of buildings not originally intended to house people.

There are examples of this kind of adaptive use in every city and state. Boston, an old city, is full of them. There are churches turned into houses or multi-unit condominiums; going beyond housing, there are railroad depots turned into restaurants, large private houses into monasteries and schools. Thirty years ago a carriage-house attached to a suburban mansion was made into an antique auto museum. Twenty-eight years later, dignified now with the title Museum of Transportation, it moved into what was

3

once a waterfront warehouse. The Boston branch of Bonwit Teller occupies a building that started life as a museum of natural history. An old piano factory has become home and studio space for a community of artists. A photographic plant is now housing the elderly. The Baker's Chocolate factory in Milton, Massachusetts, long famous for its fragrance, is being transformed into condominiums.

Elsewhere around the country there are numerous examples of adaptive use. In Soho (south of Houston Street in New York City), industrial lofts have been metamorphosed into fashionable apartments. Old textile mills in Manchester, New Hampshire, and Providence, Rhode Island, have been turned into offices and apartments. Pittsburgh and Cincinnati both boast railroad stations reborn as shopping malls. In Galveston, Texas, the Rosenberg Building (formerly commercial and loft space) has been rehabilitated into shops and apartments. "The Turrets," a mansion in Bar Harbor, Maine, is now a college hall. And I've heard of an old log cabin that was moved 476 miles from its original Tennessee site to a town near Birmingham, Alabama, to become the Brierfield Dental Arts offices.

So we began to look at properties around Boston that might lend themselves to adaptation for housing. Our criteria were few but firm. We wanted a building that we would end up transforming, at least in part, from what it had been into something else. We wanted something dilapidated but with enough structural soundness so that we wouldn't be forced to tear the whole thing down and rebuild it from scratch. And we had to find a place within half an hour's drive from WGBH.

In addition to these basic requirements, the quality of the site was important in our decision—as it would be to anyone thinking about investing in property. Television production may make special demands, but anyone must watch out for possible pitfalls intrinsic to the site. Are you on the flight path of the local airport, or within earshot of an expressway? What about the light at different times of day? Will you be able to cut down trees? You should also look at the neighborhood, its inhabitants and its ambiance. Try to find out future plans for any undeveloped areas of land nearby. And, finally, it's always worth establishing just why the property is for sale in the first place.

Bearing all these considerations in mind, my production assistant, Avra Friedfeld, began scouting around. We sent out informal requests to real-estate people, asking them to let us know what they heard. Bob Vila

4

The Chickering Piano Factory building in Boston's South End has been converted to studios, apartments, and gallery space for local artists and craftsmen.

Not many families can afford summer cottages on this scale anymore. "The Turrets," in Bar Harbor, Maine, built before the days of income or property taxes, has been re-used by the College of the Atlantic as an administration building.

spotted one good possibility, a shut-down gas station. It looked perfect: great flexibility, all kinds of landscaping potential. But before we could make a move it had been converted into a doughnut shop, and that was the end of that.

I called a friend, an urban planner and banker, who helped me meet with some builders, developers and city officials. The City Treasurer was among them, because he knows where the foreclosures are coming and where the pockets of renewal are about to happen. We got together for a free-wheeling lunch where we kicked around a whole bunch of ideas. One of the developers offered a thought: Why didn't we buy a piece of land—there are hundreds of available lots in the city—and have our head carpenter and chief genius, Norm Abram, build a house from the ground up?

This was a new angle. It looked like a smashing idea. I went as far as meeting with the Commissioner of Real Property, and she put together a list of places the city owned that had been taken for unpaid taxes and could be bought for almost nothing. As it turned out, however, we felt we were committed to the "Old House" framework for the series in question. I still like the idea, though, and some future season may well find us in a vacant lot.

Another idea we explored was to take over some surplus building from the city— an unused schoolhouse, for instance. We looked at a few, but the only one that appealed to us was unavailable—several developers were lined up ahead of us.

All this time Avra was following leads, from newspaper advertisements to letters and phone calls that came to us at the station. People would send pictures of their house or their church or their industrial property and ask whether we'd like to use it for our next series. One person even sent photos of an attractive barn and suggested we jack it up and truck it from New Hampshire to Boston. But in every case there was some insurmountable problem.

We looked at places in old sections of Boston. But one neighborhood didn't need us; they'd already been discovered and rehabilitated. Another area, Chelsea, a small city that abuts Boston, has interesting old buildings—solid row houses. It also has a sublime view of the Boston skyline. And one day, it's quite possible that "This Old House" may find itself fixing up one of those row houses, but we just never found the right one. Meanwhile the weeks went by. Autumn had set in, and there was an increasing sense of urgency. We couldn't risk losing Norm and his crews; we had to be able to

5

The Brierfield (Alabama) Dental Arts Building is an adaptive restoration of a 1790 log cabin which was disassembled and moved from Tennessee to its present location. Re-use involved adding a pre-cut cedar house to the original cabin in order to accommodate clinic space, while the cabin itself serves as a waiting room and exhibition area.

Typical of the intriguing old buildings we saw is this former Silkworm Factory in Newton, Massachusetts.

guarantee he'd have work during the winter. We had to start filming soon, anyway, if we were going to have "This Old House," Series Two, ready to air on schedule.

Our Red Elephant

The first time I saw the Bigelow House was on a sunny, blustery day in October 1979. A real-estate broker in Newton, a suburb just west of Boston, had introduced me to Mrs. Elsie Husher, the co-chairman of the Newton Historical Preservation Association, one of whose special interests was an old house there.

We drove up a fairly steep, winding lane to the top of Oak Hill. The leaves were still on the trees and the place was so thickly overgrown that we practically had to cut our way in to the house.

Oak Hill is one of a series of *drumlins*— lens-shaped or elongated oval hills of glacial drift, left behind by the departing ice something like ten thousand years ago—that make up the landscape around Boston. The site of this one is spectacular. Looking out over the hills, there is nothing between you and the horizon. You can see Blue Hill, and it's really blue. In the other direction you can see Mount Wachusett and Mount Monadnock.

The house we found there was also spectacular. It sat on a hilltop, running from north to south, a barn at the north end and in the middle a rectangular courtyard flanked by sheds and outbuildings. At the southern end, the main house stood up tall under its steeply pitched roof. ▶ The whole series of buildings made up a single, rambling, red-shingled shape, which I would soon be calling "our red elephant." Every one of what seemed to be hundreds of windows had been boarded up with plywood. Beat-up gray asphalt shingles covered the massive roofs. The exterior walls were faded and patchy-looking.

Inside the main house, we found dirty ▶ institutional-green walls with graffiti all over them, smashed toilets and bathtubs, tiles missing from around fireplaces, woodwork splintered, stairway banisters simply gone. At the bottom of the front stairway, one lonely newel post still remained. Upstairs, all the spaces had been subdivided for use as a dormitory, with a mishmash of closets and bathrooms stuck in here and there.

When we got to the sheds and the barn, however, I started to feel more hopeful. The barn had a dramatic quality, and its size and openness immediately appealed to me. Very shortly, we were agreeing that this was it, and over the next few weeks we learned all we could about the history of the place.

It had been commissioned in 1886 by a

8

This view of the Bigelow House in 1895 shows a roof covering part of the veranda under the turret as well as the original shed roof of the sunbath.

Most of the Storrows' photographs were taken around the sunbath and verandah, but this picture shows a little of the yard and planting surrounding the house, making it clear that the house was a country estate.

The sunbath roof was fitted with a moveable flap which let in light from above.

well-to-do doctor named Henry Jacob Bigelow, to be designed by the renowned architect Henry Hobson Richardson. Dr. Bigelow, by then a widower, had a formal residence in downtown Boston and also owned a shooting camp on Tuckernuck Island, off Nantucket. Now he wanted something closer to the city, where he could breathe country air and pursue an interest in botany.

Born in Boston in 1818, he was graduated from Harvard—where he had become friendly with Oliver Wendell Holmes—and had become a surgeon. He had introduced the use of ether for anesthesia, improved the

design of microscopic lenses, and devised new techniques for operating on hip joints. He had also traveled widely, established a charity clinic, raised pigeons and monkeys, played the French horn, and concerned himself—along with more common interests, including botany and natural history—with the welfare of elderly carriage horses. He married Susan Sturgis in 1847; they had one son, William Sturgis Bigelow, who also became a doctor.

The elder Dr. Bigelow lived only four years after commissioning the house on Oak Hill. In a talk delivered at a memorial gathering, Oliver Wendell Holmes remembered him at home: "There I last saw him, not really well, but not complaining, planting the young trees under the shade of which he was not to sit, looking with delight on the far-off mountains which bounded his landscape. . . . The trees are still growing, Monadnock and Wachusett are looking down upon the home he created, and he is gone."

By 1886, when Dr. Bigelow asked Richardson to do the house, Richardson too was ailing, though he was only forty-seven—a tall, corpulent, bearded, rather messy man, who liked to have himself photographed in the hooded robe of a monk. Born in Louisiana in 1839, he had grown up in New Orleans and come north to enroll at Har-

vard, and then in 1860 had gone on to study architecture at the École des Beaux Arts in Paris. Trinity Church at Copley Square in Boston, built in 1876, is probably his most famous single building. Like many of his best-known works, it is a massive stone construction notable for its low, semi-circular arches—the hallmark of a style known as "Richardson Romanesque." But there was another side to Richardson: rather late in his career he developed a type of domestic architecture built of wood that became known as the Shingle Style. The Bigelow House is certainly representative.

Houses in the Shingle Style were popular in the late nineteenth century, particularly for summer resorts along the east coast. Spacious inside, bulky and gabled and asymmetrical outside—an exterior design aimed at emphasizing the overall volume of the building—such houses were often covered with shingles from the roof right down to the ground, with a minimum of ornamentation, so as to produce the effect of a continuously flowing shape, a kind of textured mass. The style had a direct influence on the work of Frank Lloyd Wright. One of Richardson's best houses in this style is the Stoughton House in Cambridge, Massachusetts.

At the time Richardson sketched out the

10

Richardson moved his architectural office from New York to Boston when he won the design competition for Trinity Church in 1872. Completed in 1876 and dedicated in 1877, the building established Richardson's reputation and set a style for Romanesque forms in stone.

Stoughton House, Cambridge, Massachusetts (1882–83).

Low House, Bristol, Rhode Island (1887). As designed by McKim, Mead & White, the large single gable of the roof line defines the volume of the entire house, giving it a very modern look. Unfortunately, the house was torn down in 1962.

"Tippecanoe Place," the former Studebaker Home in South Bend, Indiana, is typical of the large, private mansions built in "Richardson Romanesque" style. Today it functions as a restaurant.

plans for the Bigelow house, he was chiefly preoccupied with designing the Marshall Field store in Chicago, which was to be his last major work. He died only two months after receiving Dr. Bigelow's commission, leaving his successors at the firm of Shepley, Rutan and Coolidge (principally George F. Shepley) to finish executing the design.

Even so, the Richardson touch is unmistakable—the huge, steep roofs and gables with their little lift at the eaves; the fanciful turret or "folly" at one corner, and its echo in the stable cupola; the windows set in at angles on projecting sections of the upper floors; the attention paid to the stairs, ▶ with their three different styles of fluting on the balusters; the narrow decorative line of triangular "dragon's teeth" ◀ that runs around the building and the carved ends of the lookouts over the front door; and the assorted shapes, sizes and locations of the 104 windows, sprinkled up and down and all along the east facade. The house is not considered one of the sacred Richardson monuments. It was, after all, a relatively simple summer house. But it has, in the words of his biographer, Henry-Russell Hitchcock, "distinctly Richardsonian features of the best sort."

After Dr. Bigelow's death in 1890, his son inherited the property and rented it out to

12

Dr. Henry J. Bigelow, in a photo taken in 1888, two years before his death.

Henry Hobson Richardson, in an 1886 portrait by one of his clients, Sir Hubert von Herkomer.

the family of J. J. Storrow—a co-founder of General Motors, and the person for whom Storrow Drive in Boston is named. In 1907, an art dealer named Horace K. Turner, after some negotiations with the younger Dr. Bigelow, set up the School Arts Company on the premises—an enterprise for producing and selling engravings, and for teaching students how to imitate old masters. It went bankrupt in 1912, and between then and 1919 the property appears to have changed hands several times. Finally, in 1919, it was taken over, and the main house converted into a dormitory, by the New England Peabody Home for Crippled Children—mainly

those suffering from the effects of polio and bone tuberculosis. By 1950, pasteurized milk had pretty much wiped out bone TB, and Dr. Salk's discovery would soon be on the way to wiping out polio. Meanwhile, Dr. William Bigelow had died a bachelor, leaving no heirs. The Peabody Home fell on hard times; in 1962 it finally closed, and the entire property was leased to the City of Newton.

The plan of the city authorities had been to run the place as a school for handicapped children. For a time the house was used as classroom space, then it became a nurses' residence. But when a law was passed that required the mainstreaming of handicapped students, the school closed. In 1974 the city boarded up the buildings, and the rabbits and the woodchucks and raccoons took over. The window-smashers, beer-can-scatterers, banister-ripper-offers, graffiti-writers, and brass-and-copper-pipe-extractors also helped. Invading teen-agers left the hatch to the roof deck open; through it and the 104 windows, every one of them broken, water streamed in, staining and rotting the floors and walls.

In 1975, knowing the Bigelow House was the next-to-the-last work of H. H. Richardson, Elsie Husher of the Newton Historical Commission filed papers to have it listed in the National Register of Historic Places. In April of that year, after John Howard, a Newton resident, stopped at the house to show it to an architect friend and found the place heavily vandalized, he and Mrs. Husher and Dennis Rieske, another member of the Historical Commission, incorporated themselves into the non-profit Newton Historical Preservation Association. Their first goal was to work with the city to preserve the Bigelow House from destruction; later they would address themselves to other old buildings.

Over the next few months, John Howard made himself into a one-man preservation

society. He started by mowing and raking and sweeping up debris around the house, and went on to paint some of the more deteriorated expanses of the outside. He alerted neighbors to watch for carloads of teen-agers heading up the hill and to call the police. He became well acquainted with the hill's animal population, including a very civilized skunk. And he maintained an intermittent vigil on weekends, on warm summer nights, and when there was a full moon—keeping the house safe just by his presence.

After he had pretty well tidied up the outside, John Howard discovered how the teen-age vandals had gotten into the house—through a basement window and along a heating shaft—and went in the same way. Inside, in the almost total darkness, he could see trash and wreckage everywhere. Even the radiators had been hacksawed off and carried away. Room by room, he pried the plywood off the windows so that he could see to clean up. After clearing away much of the rubble, he went to the Newton Building Commissioner with a proposal: if he, John Howard, repaired a door, would the city install a lock and give him a key? The Commissioner agreed, and so John Howard became the Bigelow House's official caretaker. He went on removing, cleaning, reglazing and reinstalling window frames; of the twenty-odd he repaired, only one was re-broken.

In the meantime, discussion of what the city was to do with the place continued. By 1977 all but five of the original forty acres had been disposed of. It was in that year that the Preservation Association acquired the house and the five acres it stood on, and began considering proposals for its use. Two years later we came on the scene and decided that it was just what we had been looking for.

Because of the sheer volume of space—10,000 square feet—I could see economically feasible ways to break it up into condomin-

17

Monumental public spaces remaining from the golden age of railroad travel provide elegant settings for more profitable public functions. This former waiting room of the Pittsburgh and Lake Erie Terminal is now serving as a restaurant.

The Walker-Stone House in Fayetteville, Arkansas, has been restored to its mid-life Victorian appearance. The original porch was built in a Greek Revival style.

The groom's quarters offered a most extraordinary view, out over the rolling countryside all the way to Blue Hill. ◀ The units in the linked buildings between the Main House and the barn would be comfortable but compact. The barn gave us total freedom to design within a space measuring sixty by thirty feet, with an enormous volume overhead. ▶ All in all, this rich variety of possibility seemed to make the project ideal for educational television.

We would have to start from scratch with the mechanical systems—electricity, plumbing, etc. We would be obliged, by our own goals and those of the Newton Historical Preservation Association, to keep intact everything architecturally significant about the exterior: no skylights, no new doorways, no moving windows around. And we would naturally have to design any changes in the less conspicuous facades on the north and west, and in the courtyard, so that they would be in close harmony with the rest. The value of the site as real estate mandated that we use the best available materials in every phase of building, from foundations to cabinet work. The fact that most of these materials were donated to us (in exchange for their televised installation or demonstration) made this kind of quality job economically feasible.

ium units of different sizes and price ranges, sharing the varied and spectacular views as well as the intimate, clustered feeling of the central courtyard.

The Main House still had many details worth preserving or restoring, and unlike many houses of the period, its scale was really not too grand for modern living. The barn and the shed areas, on the other hand— the woodshed, the icehouse, and the stable— would be completely transformed. Except for the groom's quarters in a dormer above the stable, no part of any of these structures had been designed for human habitation. We could give our imaginations free rein.

18

Many of the large homes designed by prominent architects at the turn of the century have been recycled as multi-family residences. The Patterson-McCormick House, a building in Chicago designed by Stanford White, has been used as a private school and now is being converted into condominiums.

The Morris Residences in Cincinnati are a successful re-use of a girls' finishing school which had been closed and abandoned for years. The exterior of the five-building stone rowhouse has been restored to its original appearance, and inside, twelve contemporary apartments have replaced the schoolrooms.

The Bigelow House

We knew we could expect some real problems. Zoning clearly precluded any multi-family dwellings in the neighborhood. I worried a little about how the partnership as co-developer with a part-time, volunteer organization would work. Any situation where the interests of history or culture struggle with those of development is bound to be sticky. Yet one reason we had gotten into such an enterprise was to show how such a thing might be done.

The financial arrangements we worked out were that the NPHA was to market, with our approval, the condominium units we would build. Any profits from the sale after costs had been met were to be split fifty-fifty between the NPHA and the station. The details of the project—legal arrangements, building permits, and so on—will be described in the chapters that follow. In some ways, knowing the process was to be televised made things easier; in others, it made negotiations tougher, since every move we made would be subject to the glare of public attention. But we were hopeful, because we all believed in what we were doing. We saw this as a venture that would inspire people to see the truly remarkable transformations that old buildings can undergo, and the rich possibilities they can offer.

20

The Lunar Science Institute (NASA), Houston, Texas, is located in the stucco mansion once owned by James ("Diamond Jim") West.

Chapter 2

Business Matters

I'm Bob Vila, and long before "This Old House" was even a gleam in the eye of Russ Morash, I was buying, rehabbing, and re-selling old buildings. That continues to be my main occupation. Sometimes I restore an old house, as we did in Dorchester for our first "Old House" series. Sometimes I adapt a building for new kinds of uses, as we did at the Bigelow House for our second series.

As often as I've done it, I still encounter details of design and construction that seem bewildering. A lot of what goes on in the building of a house can seem mysterious, particularly to the layman. But one of the things we've tried to do with the television shows is to lift some of the veil of mystery from the whole process.

There are some aspects, though, that retain an aura of complexity, to say the least. These are the legal and financial arrangements that have to be made before any of the actual work of construction can begin. Just as I'd always turn to an expert to lay a hardwood floor or install a heat pump, I turn to experts when it's time to give a project a firm legal and financial foundation.

A Firm Legal Foundation

The first process in any construction or development project is the initial investigation of the *title* to the property, commonly

Adaptive re-use is often a question of imagination. The Bower Schweizer Malting Factory in San Francisco is not a building which might ordinarily suggest residential possibilities, but the architects plan to knock out all the brick fill between structural piers and replace it with windows and terraces. The result, which looks in this render-

ing like an example of International Style projects from the '30's, will be a particularly pleasant and airy apartment building.

The extremely individual character of some old buildings is one of the advantages of re-using them. The University of Tampa has an unforgettable image in its fanciful Moorish building, which used to be the old Tampa Bay Hotel.

Typical of many residential re-use projects in cities is the renovation of whole blocks of rowhouses into apartments. This townhouse in Baltimore is one of 43 buildings being rehabilitated in the Reservoir Hill district.

known as the *title search*. Title refers to the rights of ownership and possession of a particular property. A *clear title* is one uncomplicated by any confusion about boundaries, past ownership, overdue taxes, or any outstanding claims by other parties that might jeopardize complete entitlement for the owner-to-be.

If you are doing a title search yourself, you should first go to the local registry of deeds and find the title records of the property you have in mind. You need to be certain that the present owner does indeed possess title, free and clear, and is thereby able to transfer that title to you. You may, however, find *clouds* on the title—an expressive term that refers to any restrictions or other parties' rights that limit the owner's title. Other problems would be *liens* against the property—claims by other parties for money owed, such as outstanding mortgages, unmet tax obligations, or unpaid debts to *mechanics*, the term for the laborers who have worked on the job.

Many potential buyers prefer to have a lawyer or a specialized firm do the title search. A lawyer or professional is trained to understand the precise meaning of the terminology in the documents involved. He knows the local laws and customs pertaining to property titles, and will be able to spot

subtleties that might pass unnoticed by a novice. A lawyer will also be alert for points such as zoning ordinances restricting use of the property which may not be spelled out in the title documents.

It is up to you to determine whether a clouded title is going to be a problem. If you decide to go ahead, clouds and all, you will probably want to take out a *title insurance* policy. If you plan to resell individual units, each must have a guarantee of clear title to be saleable, and such an insurance policy protects you (or your lender) from any loss arising from undiscovered defects in the title. Title insurance firms will require that all encumbrances be listed carefully in the policy, and will certify that the title is clear, with those specified exceptions.

The title search in the case of the Bigelow House was very clearly a job for a legal expert because of the long and convoluted history of ownership of the property, which culminated in its becoming owned by the City of Newton. The Newton Historic Preservation Association (NHPA) had their lawyer do the title search, and they also asked their lawyer to help draw up the *title deed*.

A title deed is the written document which confers legal right to ownership of property. It contains detailed descriptions of the prop-

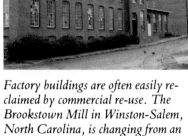

Factory buildings are often easily reclaimed by commercial re-use. The Brookstown Mill in Winston-Salem, North Carolina, is changing from an abandoned complex of textile mills to a new group of offices, shops, and restaurants.

Standout landmark buildings are obvious candidates for re-use, but any new use will have to be compatible with historic restoration. The agency which uses the Follett House in Burlington, Vermont, for its offices was able to trade off modern utility for careful restoration.

erty, along with accounts of any impinging rights or interests that the seller (or other parties) may retain in the property which will encumber the title. These encumbrances may express restrictions on how the property may be used. Or there may be *easements* (or easement rights) written into the deed—rights granted to municipal governments or local utilities to have access to parts of the property no matter who is the owner. Or there may be other *rights of way* which permit other parties or neighbors to travel across a portion of the property. Such encumbrances *run with the land*, which is a way of saying that they are transferred with the title to each new owner.

One special type of deed is called a *quitclaim deed*. A quitclaim deed grants only whatever rights or interests in the property the grantor actually has. It may not even transfer title. Or it may transfer title along with specified encumbrances.

In their efforts to secure title to the house and its five acres of land, the NHPA had worked out a quitclaim deed with the City of Newton. The deed incorporated easements for water, drainage, sewer, and power lines. Also part of the deed were rights–of–way for pedestrians from a nearby development, as well as for City vehicles en route to a nearby water tower.

Preservation restrictions were also written into the Bigelow House deed, at the request of the Newton Historical Commission—"No alterations shall be made to . . . the facade of the Main House . . . the roof profile of the Main House . . . the chimneys of the Main House. . . . The grounds of the Premises shall be maintained in a naturally landscaped state consistent with the existing topography, natural vegetation and historical use and character. . . ."

In addition, of the Bigelow House's five acres, three were subject to a permanent *conservation restriction*. A conservation restriction means that the acreage (which in our case consisted of two triangles of land, one at either end of our roughly elliptical lot) must be kept in its natural state and free of buildings, in order to contribute to the quantity of green open land in the community as a whole.

Very often a quitclaim deed such as ours will simply spell out the interests of the seller and the buyer. In some cases, the existence of a quitclaim deed will limit the usefulness of a property or its marketability. We wondered if the preservation or conservation requirements written into the Bigelow House deed would make it hard for the NHPA to market the condominium units. We decided it probably would not.

Firehouse #25 in Seattle, Washington, was in active use as a station from the time it was built in 1909 until 1973. Re-use as housing rather than office or commercial occupation was proposed because parking requirements could not be met for patrons. Even so, the architects had to excavate parking space under the existing building for owners of the 16 new apartments.

Purchase and Sale

While the title deed was being worked out, the NHPA had also been at work on another phase with legal implications: the *purchase and sale agreement*.

A purchase and sale agreement, also known as an "agreement of sale" or "contract of purchase," is a contract in which a seller agrees to sell and a buyer agrees to buy, under specific conditions. It defines exhaustively what the property consists of. It specifies all the encumbrances to the deed. It provides in detail for all of the terms of sale, down to the method of insuring the property during the time between the writing of the agreement and the actual delivery of the property to the buyer. It says what will happen in the event of natural disasters. It stipulates the responsibilities of the seller, chief among which is to deliver clear title to the property by a specified *closing date*, with no encumbrances except those spelled out in the agreement. If for some reason the seller cannot convey clear title by the date—and hour—specified, the agreement gives the buyer the right to opt out without financial penalty. The agreement is dated and signed by both parties.

Condominiums

We had wanted from the start to demonstrate how some type of multi-family arrangement could be worked out within an old building. With the Bigelow House we had an old building with far too much space for modern single-family use, but plenty for multiple units; and an old building whose exterior we were going to preserve or restore, but whose layout seemed to lend itself beautifully to multiple residences. Condominiums seemed a natural choice.

Our WGBH-NHPA team had its reasons for wanting to create condominiums. We did not want to make the place into rental units and become landlords, with all the open-ended responsibilities that would entail. We did want to set up a structure under which the units could be sold outright, with an established plan for the day-to-day management of the common areas, and provision made for the rights and privileges of everyone concerned.

Basically, condominium ownership is ownership of a living space, or "cube of air" as it has been called. You get a title deed, property tax liability (where applicable), and the advantage of income tax deductions for mortgage loan payments. You also get the advantage of *equity* that builds up over the years of your ownership of the unit.

26

Guernsey Hall in Princeton, New Jersey, is one of the earliest examples of conversion of a large estate into several luxury apartments. New owners have the collective advantage of a grand entrance and mature landscaped gardens.

This mansion in Southampton, New York, designed by Stanford White, has also been saved from demolition by conversion to condominiums. In addition to the 35-room house, there are several new apartment buildings included in re-use of the 16 acre estate.

One of the most valuable categories of re-usable real estate is large mansions which are too big and too expensive for single families to keep up. Often built of high-quality materials, like this stone house in Denver, with costly details and beautifully designed rooms, they offer exceptional opportunities for residential re-use.

Equity is the value of a homeowner's unencumbered interest in a piece of real property: that is, the property's market value less any unpaid mortgage balance or any other liens or debts for which the property is security.

As a condominium unit owner you get a share of the *undivided interest* in common areas such as driveways or lawns.

As recently as a generation ago, condominiums were rare in this country, although they'd gained wide acceptance in Europe. Puerto Rico started the condominium boom; next came Florida; and other states up the eastern seaboard and then across the country soon followed. Now, in some areas at least, condominiums account for twenty to forty percent of available housing units. In other areas, they constitute as much as fifty percent of housing starts. HUD has predicted that by the year 2000, fifty percent of Americans will be living in condominiums.

Why the boom? There are great advantages in owning a home where someone else mows the lawn, maintains the roof, pumps out the septic tank, rakes the leaves. It is an attractive combination, and if it's managed well it can be eminently satisfactory.

There are, to be sure, other types of home ownership involving multi-family organizations. One form is the *cooperative* apartment, or co-op, which is most familiar in New York and Chicago. In a cooperative apartment building there is just one title deed to the whole building. The co-op dweller purchases shares in a corporation which owns the building. Each shareholder then has an exclusive right to occupy his or her apartment. There are some distinct differences between a cooperative and a condominium. For instance, in a co-op there is real control over who the inhabitants will be; this isn't true in a condominium. In a cooperative you work much more closely with your neighbors. Your financial fate will also be closely linked with theirs, as property values, tax assessments, etc. rise or fall.

Another approach to ownership is the *tenant-in-common* structure. In this form of ownership all residents together own every part of the building. Each tenant-in-common has the right to occupy any part of the building. The details of who pays what on taxes and upkeep and mortgage loan installments, and who decides on improvements or changes of ownership, have to be thrashed out case by case. This kind of arrangement demands enormous mutual cooperation. It's definitely for people who get along well together.

A condominium, by contrast, explicitly distinguishes individual from common property. Every condominium functions

On a more modest scale, large Victorian homes also offer enough space for more than one family. This rambling, two-story house in Princeton, New Jersey, is now divided into five apartments.

Other re-use projects are a triumph of ingenuity. This store-front and former Christian Hall in Crested Butte, Colorado, is now a health club in one of the West's best skiing areas.

under a document which is often called a *master deed*. Other typical names for this document are "declaration of conditions, covenants, and restrictions" or "plan of condominium ownership." The master deed defines the condominium as a whole and establishes the precise boundaries of each unit. Here is some of the wording from a fairly representative master deed I studied recently, just to give you an idea: . . . *The boundaries of each of the Units with respect to the floors, ceilings, and the walls, doors and windows thereof are as follows:*

(a) Floors: The upper surface of the subflooring.
(b) Ceilings: The plane of the lower surface of the ceiling joists or, in the case of Units or portions of Units situated immediately beneath an exterior roof, the plane of the lower surface of the roof rafters.
(c) Interior Building Walls: The plane of the surface facing each Unit of the wall studs.
(d) Exterior Building Walls, Doors and Windows: As to walls, the plane of the interior surface of the wall studs; as to doors, the exterior surface thereof; and as to windows, the exterior surfaces of the glass and of the window frames.

This Master Deed is equally exact in defining the common areas: *The common areas and facilities of the condominium consist of:*
(c) the entrance lobbies, halls and corridors serv-ing more than one Unit and the mail boxes, closets, fire extinguisher and other facilities therein, stairways, fire escapes, and elevators . . .
(e) all conduits, chutes, ducts, plumbing, wiring, flues, and other facilities for the furnishing of utility services or waste removal which are contained in portions of the Building contributing to the structure or support thereof.

I've quoted these passages just to give a sense of the minute and careful wording of the typical master deed. Other areas covered in the master deed may include how the units may be used (residential use, business use, etc.); how many persons or family groups may inhabit a unit; what type of trust or association of owners will be in charge of administering the condominium; how legal enforcement and insurance coverage have been arranged; and by what formula the unit owners' *interest percentages* have been computed.

This last item, the individual owner's percentages of interest in common property, can be established under any number of formulas, according to state laws or local customs, or both. One owner's interest may be in proportion to the original appraised value of his unit as a fraction of the value of the condominium as a whole. Or it may be in proportion to his square footage as a frac-

28

"The Castle" is a Victorian neighbor of the Bigelow House, located on another drumlin in Newton, Massachusetts. It was renovated and converted to condominiums as the pioneer development under a new zoning ordinance which permits multi-family re-use, thus avoiding institutional development, demolition, or subdivision of the property.

Many historic buildings are no longer used for their original purpose. This old temple in Cambridge, Massachusetts, no longer has a congregation but is being recycled as studio space for artists.

tion of the total square footage; or to the market price of his unit as a fraction of the total of all the market prices. Or in a condominium consisting of many closely similar units, each unit may simply have an equal share.

What is the importance of the unit's interest percentage? It's considerable. It sets the unit owners' relative voting strength in matters brought before the group. It is directly related to the amount of the unit owner's monthly maintenance bill (which in turn goes to pay the costs of running the building).

The creation of a master deed is yet another process where legal help will be important; and it's the responsibility of the developer to round up the lawyers and write the deed. This also holds true for an accompanying document, the *bylaws*. The bylaws deal more with procedures than with definitions. How will voting be carried out? How will responsibilities for repairs be decided? How can the bylaws themselves be amended by the owners once the condominium is established? All these questions must be addressed before a single unit is put up for sale.

I could see some interesting challenges ahead for the new co-venturers, WGBH and the NHPA. It was true that a lot of the charm of this project for television purposes was the wide range of sizes and styles of units we could envision. But this could create problems. Would the Main House rule the whole roost, simply by its disproportionate size and price? Would the wide discrepancies in income levels required to support the different units mean equally divergent lifestyles and outlooks among owners?

We decided we'd just have to address these matters in their own good time. Meanwhile, we had work to do. An early item on our long agenda was a *zoning variance*.

Aldermen and Ordinances

Zoning restrictions are established by every city or town for every section of the locality, and in a densely populated community they can get pretty elaborate. The first thing that a would-be builder, developer, or rehabber should do is clear up any conceivable zoning difficulties.

I've seen some terrible examples of what happens when zoning issues are not confronted at the start. Eight or ten years ago, in my own area, a fast-food chain built a restaurant on a lot zoned for residential or retail but not restaurant use—and the neighborhood rose up and stopped them dead. The building has stood there ever since, boarded

29

The Day-Taylor House had been scheduled for demolition as part of urban renewal in Hartford, Connecticut. When city plans were revised to encompass rehabilitation instead of rebuilding, the house was repaired and restored. Today it is used for offices and one residence.

The elegant hammered metal facade of this factory building in Detroit was once an advertisement of the company's craftsmanship. When the operation moved to larger quarters, new owners saved the building by putting in office, studio, and restaurant space.

over, surrounded by tall weeds.

Very often, the zoning research and any request for a variance is another of those processes best handled by a lawyer. Lawyers know where to go for information and what steps to follow to request changes in zoning. As they research the zoning and the community in general, they're able to determine if the city or town has any yet undisclosed plans for your site. No one wants to build in the path of a planned freeway or in the middle of a power plant that's still on somebody's drawing board.

Our project at the Bigelow House presented zoning problems. Before the NHPA had become interested in the land, the former Bigelow-Peabody School property had been unzoned. The Newton Board of Aldermen finally zoned it as Residence E. Residence E provided that on this land no building could be constructed to house more than two families. On the other hand, the ordinance did state that petition could be made for (among other things) permission to build a multi-family structure.

Such variances are always dealt with on an individual basis. They require a series of meetings complete with advance notice and a good deal of paperwork. Planning boards, appeals boards, conservation commissions may be involved. If all moves smoothly, the

process takes four to six months. It is very much a public process, and as with many other aspects of this preparatory phase, careful homework is important.

Your responsibility is to demonstrate convincingly to the community that what you plan to build (or rebuild) will be a desirable addition for the neighborhood you're in—whatever the zoning says. Naturally enough, this is most tricky when you are intruding a non-residential element into a quiet residential area, or when you're proposing a thick population cluster in an area of widely spaced private homes. Here's where the condominium concept is beginning to make real inroads into old established patterns. More and more often, appeals boards are feeling that a structure housing, say, twenty-one families, surrounded by open fields or wooded land, offers better amenities both to its residents and to the community at large than twenty-one individual houses with individual yards and individual carports.

Our lawyer helped research and present our case; so did our architects; and so did the NHPA, who had in fact already taken steps to get a variance for a four-family proposal. With all this ammunition and with the promise of a fine new sum of property tax revenue for the City, it wasn't hard for us to

In a factory district of Chicago, this little building which once contained a cookie factory has been resurrected as a small apartment house. In this case, rehabilitation and conversion to condominiums adds to the city's housing stock, while halting the neighborhood's decay.

show that our condominium plan would enhance the beauty of the hilltop and the desirability of the whole neighborhood.

Nevertheless, the procedure did indeed take six months. The twenty-four member Newton Board of Aldermen met. We presented our drawings and our descriptions. And in the spring of 1980, the Board of Aldermen granted a Special Permit and Site Plan Approval.

Appraisal Time

Long before the zoning variance was actually in our hand, we had an *appraisal* done on the Bigelow House and land. Together with the zoning variance this appraisal was needed to help support our case when we went to look for financial backing.

Appraisal is both an art and a science. A good appraiser has to have experience in real estate and in the financial world. There are firms that specialize in appraising; there are also many real estate firms that do appraising along with other aspects of the business. From the developer's angle, the important qualifications of an appraiser are:

• An impeccable reputation for care and accuracy.

• Solid professional credentials.

• Extensive local knowledge and an ability to predict based on past history and future trends.

The appraisal process is not anywhere near as time-consuming as the zoning variance process. Usually, however, it is the first major expense you face. You may be able to get an appraisal done for as little as fifty dollars. The more complex the proposed project, the more elaborate the appraisal—and the fee can add up to more than $2500.

There are three ways to compute an appraised value.

• The *cost approach* bases itself on the actual recorded cost of building a new or nearly-new structure. What did materials and labor and all the other costs of construction add up to? That sum constitutes the appraised value under a cost-approach appraisal.

• The *income approach* endeavors to compute projected annual income if the property is to be rented or used for business purposes. This income will govern how much the owner will be able to spend over a period of time to repay lenders, improve the property, pay taxes, and so on.

• The *market approach* evaluates the per-square-foot dollar value of the property

31

In Galveston, Texas, this historic brick and cast-iron commercial block has been restored and re-used as city apartments above store-fronts.

This two-story brick building in Denver was built to house railroad workers over commercial manufacturing space on the ground floor. Renovations in 1972 introduced office space on the first floor and ten townhouse-condominiums on the second, with a tenants' tennis court and pool.

when developed as planned and placed for sale on the open market.

The market approach is the one we used, because we did plan to put the units for sale on the open market. Here again, for all its oddities, the Bigelow House appraisal is a perfectly good sample case. We located an excellent local real estate firm. We told them what we needed. Their report contained the following:

- A letter of transmittal stating the total appraised value of the projected condominium development as a lump sum.
- A list of *assumptions* and *limiting conditions*. For instance, "that the title of the property is marketable"; "that no responsibility is assumed for legal matters"; "that normal access to the site is a prerequisite for the estimate of value cited herein"; "that the value . . . represents the value of the property in the ordinary course of business and under stable conditions and not on a forced liquidation basis"—and so on.
- The qualifications of the appraiser himself; education and training, licenses, board memberships, teaching positions held, past experience.
- Photographs of the exterior and interior of the Bigelow House as it was at the time, and views of the site.

- A statement of the purpose of the appraisal, "to estimate the market value . . ." and a definition of *market value*: "The highest price in terms of money which a property will bring in a competitive and open market" under straightforward conditions at the present time.
- Definitions of other terms used in the report.
- A brief history and profile of the City of Newton—how it grew from thirteen scattered villages, how the diversity and character of the villages remains in the neighborhoods, of which Oak Hill is one—and some statistics about City population, industry, business.
- Data about the Oak Hill neighborhood itself with its many single-family residences and nearby junior college.
- Zoning data, tax assessment data, and a map of the neighborhood.
- A statement that condominium conversion would be the *highest and best use* of the property; that it would be "that usage which returns the highest yield to the land and is still in conformity with the neighborhood usage."
- A history of condominium developments in Newton and nearby communities.
- Data about the Bigelow House site—its physical properties, its water and sewer

32

In Little Rock, Arkansas, an architectural firm bought this Greek Revival home—built for a prison warden with prison labor and prison-made bricks—and renovated it for use as their offices.

lines, its outstanding variety of native, European, and Asian shrubs and trees.

- Data about the house itself, with indications of the proposed re-use of the different parts of the house and outbuildings.
- A study of four comparable developments in greater Boston. Some were more comparable than others, but in each case an old existing structure had been adapted and/or enlarged to house new condominium units. This study included a table of per-square-foot valuations for all the developments, adjusted for such tangible and intangible factors as inflation and location and appearance.
- An enumeration of the values per square foot of all the Bigelow House condominium units (subject to their being built under the budget and with the materials proposed) which yielded the grand total named at the beginning of the Appraisal Report.

There is also a page of Certification, signed by the appraiser, where he warrants that this is a truthful and disinterested appraisal and "subject to . . . the Code of Professional Ethics and Standards of the American Institute of Real Estate Appraisers of the National Association of Realtors."

Following the main body of the Report are fifty pages of addenda and exhibits.

Among these are the agreement under which the NHPA had obtained the land; the deed to the property; zoning ordinances and variances; more photographs of the area; a list of Oak Hill flora; architects' plans and elevations of the condominium layout; the estimated budget for the project; maps of areas where comparable condominiums were located, and photographs of those condominiums inside and out.

I have given a blow-by-blow description of our appraisal because this seems to me the clearest possible way to show the kind of groundwork that a hopeful adaptive user should be prepared to undertake. I should also say that an appraisal report is a very good read. The outlook of the appraiser comes through, as do the desires and prejudices of neighbors and city officials, and the energy and imagination of other comparable developers.

No matter how grand the qualifications of the appraiser or how methodical and mathematical his exposition, the fact is that, to some degree, he has to go by something like intuition. He presents his report on the clear understanding that it is just an educated guess.

The same applies to other kinds of appraisals you should be thinking about. During the same time period you need to get a

33

A former coach house in Andover, Massachusetts, has been converted to a modern residence with very little exterior alteration. The addition of new window openings and relocation of a chimney to serve a new living room fireplace are the only visible exterior changes. Garage doors have been added to close off storage space under the house.

physical inspection of the premises by an engineering consultant or inspecting firm; and you need to start figuring out your costs in detail.

How to Think Like a Banker

As we prepared ourselves to meet with our banker, more correctly called *lender*, the appraisal was able to furnish us with positive selling points. The negative side of things must be laid out just as carefully—in particular, the projected cost of what we planned to do.

First we had our inspector prowl around the house and poke at the foundations. The bank was having their appraiser make his best guess at the market value of the condominium development. They also sent their structural engineer to check the soundness of the existing building.

As it happened, both inspectors fell very far short of guessing how much reconstruction we'd have to do. At least they agreed with each other, though, which certainly simplified things at the time.

Having an inspection done isn't a very costly part of the process. The inspection fee may be just $50 or $100, and in most cases it is money well spent. In addition, many inspecting firms certify their results. This means that they not only sign their name to

their educated guess, but carry insurance in case you sue them for steering you wrong.

Our old building had a lot of hidden flaws that were unprecedented in the experience of our inspector. Even Norm Abram, an exceptionally knowledgeable rehabber, was caught unawares. He was more pessimistic from the start than the consulting engineers, but not pessimistic enough.

Inspection reports in hand, we started to put together a budget. The budget is another perfect example of educated guesswork—at this stage, anyway. We aimed to do two basic things—to judge what fraction of our overall costs must be allocated to each individual operation or process or material, and to make as close a stab as possible at estimating the heaviest possible cost of each item and each operation.

One incident suggests just how tough an assignment this is, even for very experienced builders or developers. Norm Abram and I sat down to make some lists of worst-case figures for various things we hoped to do. In areas where Norm knew more than I did (like masonry work and carpentry) his estimates were double mine. In areas where I had more background than Norm (like paving and landscaping) my estimates were double his.

The lesson to be learned here is a sobering

These twin hotel buildings in Oak Park, Illinois, have been gutted and renovated as apartments. The architects roofed the space between the wings and transformed a bleak alley into a bright interior atrium.

one, but we all have to learn it. The important point to make now is that the better informed you are the more realistic you can be; and the more realistic you can be, the better. It will be for your own good in the long run. Discouraging as it may be to come up with a high cost estimate, it is a lot more discouraging to run out of funds before you finish.

Staying fully informed all the way through is absolutely vital. A part of our budgeting process that began before we even had an appraisal, and that continued until the project was completed, was the monthly update of estimated and actual costs of every single item and operation that went into the job. By faithfully tracking costs this way, we could immediately spot anything that was starting to get out of control. Where there were major over-runs, we were able to make decisions on cutbacks or sacrifices in an attempt to balance things out. At the same time, we were able to keep a wary eye on the small but insidious upward creep of everything.

Sometimes our budget seemed to be sprouting whole handfuls of sore thumbs. But at least knowing about these contingencies quickly helped us handle them more efficiently.

A good method of tracking costs is to use the Contractor's Application and Certificate for Payment forms of the American Institute of Architects: AIA Document G702A, for short. A sample G702A from a job near completion (a good deal smaller and simpler than the Bigelow House job) showed extremely careful cost control—despite a couple of change orders that had made their way in. The *retainage* column represents the portion of total billings (this portion varies, but ten percent is an average amount) that is retained by the owner, as agreed between owner and contractor, and not paid until a certain number of days (perhaps thirty or forty-five) after the job is completed. Retainage serves as a way of insuring that the job will in fact be completed, since it usually constitutes a large part of what will eventually be the contractor's profit.

For accuracy and helpfulness, it is wise to make these lists as ample and detailed as possible. And these forms, or others like them, should travel back and forth between developer and lender with great regularity. The relationship between developer and lender is a partnership, with risks and opportunities shared by both partners. The more comprehensively informed your banking partner is, the more effectively he can play his part.

35

This historical firehouse was sold by the city of Cincinnati on condition that the new owner rehabilitate the building. Without self-consciously exploiting firehouse trappings, the architects made use of the large coach room with its original columns, arches and piers as restaurant space and converted the upper floor to offices.

Our New Partner

The first overtures to a lender who might help finance our project had been made months before we actually sat down to come to terms. Next came the drawing of plans, the meeting with zoning boards, the making of appraisals, the estimating of costs. And as I've mentioned, while we compiled data at our end, the bank was busy getting some figures of its own, including both cost estimates and an appraisal of eventual market value.

We were trying to get a construction loan. The Bigelow House and its five acres were already in the possession of the NHPA, thanks to the purchase and sale agreement drawn up in 1977 and finalized in 1980 between them and the City of Newton. So we and the NHPA, as joint venturers, had the house to offer as security for our loan. The lender would be dealing with just one of the joint venturers, namely the NHPA—a corporation that was virtually penniless. But we also had the corporate entity that is WGBH, whose reputation and substance would serve to guarantee the loan.

On top of these rather mundane considerations, we had going for us the exciting prospect of a television series that would not only publicize the neighborhood and the

historical interest of the site, but would go very far toward selling the condominium units without any other sales effort.

We gave the bank a *mortgage* on the house. Often we speak of mortgages as though they were sums of money, but actually a mortgage is a lien, or claim, on property. Since we were using property as our security, we went to a savings and loan association rather than a commercial bank.

For a big developer, a more standard arrangement might have been an unsecured loan from a commercial bank. An individual, on the other hand, might have "bought" a place with a lender's money, giving the lender a mortgage which would be paid off, with interest, over a period of ten to thirty years. What we did was a somewhat unconventional cross between the two, because our construction loan was for a duration of no more than a year, but it was secured by a mortgage on the house.

The loan application process can take six months. As the lender looks at a loan application, he keeps in mind the appraisals and inspection reports he himself has received (and decides whose figures are closer to reality). He must have before him all relevant deeds, zoning variances, building permits, architects' drawings, survey reports, and so on.

One friend explained his bank's attitude this way: "The lender," he said, "wants to be the last fellow in and the first fellow out." He wants to help make the project work but he must stay well back of the line of fire. Above all he wants to get his money out. No bank or thrift association has the slightest interest in winding up short of money and in possession of a partly-built condominium.

So the lender attempts to decide, based on all the facts and figures you and he have mustered, to what extent he wishes to participate in the risks and the potential opportunities your project offers.

Some standard percentages come into play here; so standard that they sometimes seem to amount to self-fulfilling prophecies. The developer hopes that his costs will not add up to more than eighty percent of his ultimate selling price. Of those costs, he hopes that a lender will be willing to put up at least eighty percent, or, say, somewhat over sixty percent of the hoped-for resale price.

The lender will also require, in most cases, that a *contingency fund* be set aside somewhere (and that proof of its existence be given) as a way of hedging against total disaster if costs get out of hand. The smaller the project the larger a percentage of the total cost the contingency fund is. A maximum

37

With the trend to centralized education, many local schoolhouses have been abandoned and are available for re-use. This handsome brick school in Lexington, Massachusetts, with many "Richardsonian" forms, has been converted to apartments.

percentage would be about 20 percent.

For example, to pick a set of numbers, if you were to see probable costs of *about* $80,000 (say not quite 80 percent of the appraiser's total value of $100,000 for the finished condominium). You would probably be able to raise a construction loan of $64,000: 80 percent of the $80,000 cost estimate. And you would set aside a contingency fund of *about* $20,000.

Badly stated like this, the numbers look rather large. Eventually you learn to feel comfortable with figures that might once have seemed totally unreal. It's not a question of absolute dollar amounts but of relative scale. To the person planning a ten-thousand-dollar improvement, a five-thousand dollar overrun would be shattering to contemplate. But a million-dollar overrun wouldn't particularly faze the builder of an eight-million-dollar shopping mall.

There are any number of other, smaller figures that entered into our financing picture. *Closing costs*, as they are called, included all the fees we'd paid to appraisers, surveyors, inspectors, and attorneys; also the cost of documentary stamps; also escrow fees. An *escrow fund* is another form of contingency reserve, deposited in a special account to meet periodic expenses such as insurance premiums, taxes, and special assessments.

Our escrow fund was established as part of the procedure that would culminate our work: the closing.

Another most important matter covered at the closing of our loan was the lender's agreement to offer financing to a percentage of the future buyers of our condominium units. Financing is a central consideration for a condominium purchaser, as it is to any home-buyer, and making sure in advance that it will be available is the developer's responsibility.

Your lender will usually agree to finance a minority portion of your units—provided that you can demonstrate that more than half of them are as good as sold. You have to be able to produce letters of intent from your would-be buyers, and to show that they have put down deposits or given some other concrete evidence of their intent. But the unit buyers have to do all this before financing has been lined up, so that financing can be lined up. This can be a tricky situation. It is another reason why any condominium developer must have a close working relationship with his lender. That relationship must in turn be the basis for relationships with other banks or thrift associations that will provide financing for the purchasers of the balance of the units.

One element that is often part of a real es-

A Texas savings bank has restored a number of historical Texas farm houses for use as its branch offices. Typically decrepit, the Aynesworth-Wright house in Austin (shown above before and after renovation) was saved from complete deterioration by the adaptive re-use project.

Twenty apartments are being made in the former St. Mark's Church in Brookline, Massachusetts.

tate transaction has been missing from my account: the real estate agent or broker. In our particular case there was no broker involved, and we had enough combined experience and expertise to find our way through the legal and financial processes on our own. A broker often is, however, the pivotal contact between parties at the outset. He or she may well be right in the thick of things—informing, negotiating, translating documents for the uninitiated, all the way through to the closing date.

And In Closing . . .

The closing of a deal is just what it sounds like: the final and formal signing of all the papers, the shaking of all the hands. It is the finish line of the preliminary stage.

Our closing was quite an event. It took place at the office of our banker, and everybody was there: the WGBH people and our lawyer; the NHPA people and their lawyer; a representative of the City of Newton and the City's lawyer.

The *closing documents* made up a fat book. They ranged from copies of the title deed and certificates of the legal existence of the NHPA and WGBH, to specific terms of our construction loan. They included the terms of the mortgage the NHPA was granting to the lender; the guarantee WGBH was pro-

viding to the lender; the zoning variances and building permit the City had granted to the co-venturers; the joint venture agreement itself; the agreement with the architect; and all sorts of safeguards and commitments and binders. In spite of a lot of rather special aspects to our case, it was a fairly typical gathering of closing documents.

We and the NHPA and our lender signed a total of eight documents that day. And now at last we were really official.

The Lone Star Brewing Company buildings in San Antonio will be born again in 1981 as the San Antonio Museum of Art. The main structure—the original brewhouse—will become museum exhibition space, and the six ancillary buildings will be renovated for offices, shops, storage and restaurant.

Chapter 3

Getting Started

Now that the house was actually ours we began to feel really excited about planning the details of the design. This is always an absorbing part of the process, and even the most case-hardened rehabilitator feels a certain sense of euphoria—somewhat like a sculptor starting work on a raw block of granite. Many ideas will be discarded eventually as either impractical or too expensive, but the process of trying them out can be inspiring. Ideas generate other ideas and, in the end, the best ones will find their way into the final design.

Our raw material was an eighty-five-year-old house of historical significance and distinctive design in unusually neglected condition—particularly the interior. Most of the salvageable historical fabric was on the outside of the building. The original design took the form of a tall, but not over-large, three-story house with a lofty, steeply pitched, wood shingle roof, and red-painted wood shingle walls. A corner turret on the second floor added a touch of whimsy and romance. Richardson was perhaps influenced by the French chateaux he saw as a student, or by the Scottish castles that figured in Sir Walter Scott's novels, which were very popular at the time. Two other distinctive features were a small greenhouse extension at ground level (usually referred to as the *sun-*

Crathes Castle in Leys, Scotland, is typical of strongholds built to withstand the religious and political turmoil of the 16th century. Corner turrets, often with separate staircases, provided private and convenient locations for look-out and defense, temporary imprisonment, and romantic encounters.

Turrets were more stylish than defensive features on French châteaux, and by the time they appeared on late 19th century American mansions, they represented pure fashion. The Croke-Patterson-Campbell House in Denver is an excellent example of Loire Valley style which has been saved from demolition by sympathetic re-use.

bath) and the *belvedere*, a platform on the roof surrounded by a railing, with a commanding view of the countryside. At second-floor level, part of one of the gables had been slightly extended over the principal entrance, giving an oddly romantic shape to what turns to be a bathroom. Behind the main house stretched a lengthy tail of double and single-story units—an *ell* with a steeply pitched roof followed by the lower-roofed woodshed, icehouse, stables and barn that were grouped around a courtyard, the whole series forming the shape of a small letter *b*. Together, these other structures covered about six times the floor area of the main house—a clear case of the tail wagging the dog! The interiors of the units gave us a variety of interesting spaces as well as a number of fascinating details that were to be incorporated into the rehabilitated project.

Although it is a good idea to employ an architect for any rehab project, if you are creating more than one living unit the services of an architect become mandatory. Sometimes it is wise to consult an architect before you purchase a building to judge its soundness and decide whether or not it can be subdivided into the number of units required.

Finding An Architect

It is important that you find an architect who possesses the rather special blend of creativity, reticence, resourcefulness, and practical knowledge needed to adapt a good existing building. It is not always easy to find such a person—many architects who excel in the design of new buildings have been known to produce second-rate results when faced with the relatively severe limitations of working with something of quality that already exists. We were fortunate to have worked in the past with two trusted professionals who had experience with historical buildings on the island of Nantucket, where shingle-clad buildings not unlike the Bigelow House abound. These architects were John (Jock) Gifford and Bill Rich of Design Associates.

Otherwise, the best line of approach is to look around for good examples of work similar to what you want, find out the name of the architect and, if he or she is conveniently located, arrange a meeting. No commitment is necessarily made at this point; any reputable architect will be glad to show you examples of his work, discuss anything from the philosophy of design to fee scales, and refer you to satisfied clients. The chapter of the American Institute of Architects nearest you should have a list of architects experienced in rehabilitation and adaptive use,

and may even have the addresses of buildings they have adapted. Whatever you do, don't grudge a little time spent in shopping around at this point.

The services of an architect may be hired either at a flat fee, based on a percentage of the construction cost of the work, or at an hourly rate up to an agreed maximum. For a rehab job, where the extent of the work to be done is often uncertain, to say the least, the second arrangement is generally better and it is the one we used for the Bigelow House. Instead of a contract, we drew up a simple letter of agreement, signed by both parties, stating the hourly rate and the price ceiling under which the work was to be completed. If unforeseeable circumstances made it appear likely that the ceiling would be exceeded, the agreement could be re-negotiated. One advantage of the hourly rate method is that if you are dissatisfied with the work, you can call a halt at any time, and be liable only for services up to that point instead of for the whole job. Many architects also prefer this method, as it can protect them from the dishonest client—if not paid on a regular basis as agreed, the architect can simply stop work.

It is of the utmost importance for you to feel comfortable with the architect you choose and that he or she listens carefully

42

Architectural response to rehabilitation can range from careful restoration of period details to stylistic eclecticism. The owners of "Glencairn," an 18th century manor house in Princeton, New Jersey, have personally restored the house to its pre-revolutionary elegance. Inside a brick factory building on Printers' Row in Chicago, one architect has used the brick shell of the original as a foil for his new construction, with equally striking effect.

and sympathetically to your requirements. A good adaptive use design is usually the result of a good dialogue between designer and client, and some flexibility on both sides is essential. For your part, always try to make your requirements reasonable and state them clearly, as nothing can be more frustrating to an architect or more expensive to you than having to change drawings at a late stage in the design.

A last word of caution: avoid the architect whose work seems to consist mainly of Strong Statements—such an approach can be disastrous for older buildings! The architectural profession is just emerging from a period when, in new work, boldness and contrast with the surroundings were more highly thought of than any subtler effort to achieve symbiosis. As contrast is usually much easier to achieve, the temptation was to use older buildings simply as backdrops for bold and incongruous additions, extensions and remodeling. On the other hand, the timid architect is unlikely to come up with the resourceful and imaginative ideas often needed when adapting a building—hence the importance of finding the architect with the right balance of talents to produce a design that functions effectively while preserving or enhancing the existing building.

Measuring What's There

The first thing Jock and his team did was to measure the building completely and record the results. This is an essential step in all rehabilitation work before any serious design proposals can be made. As this had to be done far from their office and in the heart of winter, they set up a "hot room" in the house itself, consisting of plastic tarpaulins and a space heater.

Although the measuring of a building is generally done by an architect, if you have a small rehabilitation job you may want to try to do it yourself. All you need is a measuring tape, preferably about fifty feet long; someone to hold the other end of the tape while you measure; and a pad of paper on which to sketch the outline of the rooms and to record the dimensions. The sketch need not be strictly accurate so long as it clearly shows all the features of the *coastlines* of the walls in proper sequence—the things that protrude, such as chimney breasts and pipe castings, and the things that recede, such as windows, doors and alcoves. Be sure to leave enough room to write in a dimension for all of these surfaces. The sketch should be completed before measuring begins so that it will not distract you from reading the tape accurately. Although the sketch need not show the actual thickness of all the walls and par-

43

The design of the formal entrance to this barn-turned-residence is particularly sympathetic to the forms and materials of the original buildings, and yet it clearly indicates a new addition and organization.

titions, you should keep the thickness in mind if you are doing more than a one-room layout—particularly if they seem to be either greater or less than average. (An internal partition that is much more than five inches thick very probably contains pipes and other important conduits that may have to be retained in the new layout.)

It is also important to record the main dimensions of the rooms and, if a whole floor of a house is being measured, the main internal dimensions of the house itself. Without these overall dimensions it is all too easy to get lost in a maze of small ones that don't seem to want to add up, so that extra visits to the site have to be made at a later date. If you have to travel any distance or must get permission to enter the building it can be a nuisance. Often, too, buildings are not quite square, and so if there seem to be any significant wanderings in the directions of the walls, diagonal measurements should be taken from the corners of the suspect areas. If the diagonals in a room are not approximately the same, it is not truly rectangular and will have to be plotted with extra care. Fortunately for the amateur measurer, this does not happen very often, but is found occasionally in urban situations where buildings are built right up to a streetline that is not at right angles to the walls.

After all the measurements have been taken the next step is to draw the whole layout at a given scale. A scale of one-quarter inch equals one foot is handy for floor plans, where one-quarter inch on the drawing equals one foot of the actual building. However, if only a room is being drawn, a scale of one-half inch equals one foot may be preferred in order to study, say, a furniture layout more closely.

For anyone who has never drawn anything to scale before, this is the trickiest part of the process. Some people prefer to use squared graph paper, but this can become complicated if more than one room is involved, as those five inch partition walls seem to throw everything out of gear. To get good results, the best way is to use the same tools an architect would—a t-square, triangle and architect's scale—on either plain white drawing paper or tracing vellum. If the drawing is done with pencil or ink on the tracing vellum it will be reproducible by an inexpensive printing process.

Preliminary Design

Before we joined forces, the Newton Historic Preservation Association had hired an architect, Robert Lauricella, to prepare a feasibility study of converting the Bigelow House into four apartments or condomini-

This view of the stairhall of the Stewart residence in San Francisco shows how many details typically are missing in an old house. Before rehabilitation began, someone had to measure all the heights, depths, and widths in order to locate the old switches, fixtures, windows, doors, and every change in surface.

ums, mainly as a demonstration that the building was a viable candidate for rehabilitation. After re-examining the Lauricella plans, Jock and his crew found the basic subdivisions of the proposal to be well conceived and we decided more or less to go along with them. We would, however, be adding an extra unit, and, because of our somewhat different requirements, the layouts of the units were almost completely redesigned.

Design Associates now had the task of coming up with a workable layout for the five condominium units that would do as little violence as possible to the building.

The task was made a little simpler for them by the scarcity of interior historical detail, other than in the Main House and the ell, that either existed or could be re-used for residential purposes. From the viewpoint of historical restoration, the outside of the house would receive the most attention.

Working Drawings

For any but a very small and simple rehabilitation job, the law usually requires that a drawing and/or specification be submitted to the local building department before a building permit can be obtained. If the job entails, for example, the conversion of a sin-

45

gle dwelling into more than one family unit, the working drawings must usually be prepared and stamped by a registered architect. Apart from such legal requirements, however, a set of proper working drawings is needed for building permits, changes of occupancy, zoning appeals, and the approval of historical commissions. They may also be required for dealings with banks and real-estate agents, and for obtaining realistic bids or estimates from builders. They can help to expedite the work by showing clearly how it is to be constructed; and, finally, they serve as a record to future owners of the work as built. Unfortunately, no such record of the original Bigelow House had survived. We intended to make sure that this did not happen again by encapsulating a complete set of *as-built* drawings somewhere within the structure after completion of our work.

Typically, a set of working drawings will consist of plans (what you see when looking down on the layout), sections (what you see if you slice vertically through the house), elevations (what you see looking at the house from the outside), and details, which may be larger-scale sections through windows, doors, eaves, etc. For rehabilitation work these may not all be necessary. If no changes are being made to the outside, for instance, no elevations are needed. Sections,

too, are often omitted unless new vertical spaces are to be opened up inside the house. The plan, however, is an essential if there are any changes to the layout of the existing building.

In a good rehabilitation plan, all existing walls, partitions, doors, windows, chimneys, etc., that are being retained in the new layout will be outlined with a solid line. All existing walls, partitions, doors, windows, chimneys, etc., that are to be removed will be indicated with broken lines. All new walls, partitions, doors, windows, chimneys, etc., will be clearly differentiated from the existing work. Enough dimensions will be given to locate all new work in relation to the existing walls, partitions, etc., and the location of all new electrical, plumbing, and heating services will be shown.

Clear working drawings that are intelligently organized and easy to read can save a lot of time and frustration on the site; they are also a good indication of the competence of your architect.

The Local Authorities

Once we had a fairly firm idea of what we wanted to do with the house, and before our working drawings were completed, we began to approach the local authorities. Because of the historical significance of the

46

The sectional drawing (right) of this former schoolhouse in Lynchburg, Virginia, while not a construction drawing, shows how the architect opened up the center of the building to create an atrium between the front and rear apartments.

Bigelow House, we decided to apply first to the Historical Commission. Around the middle of January, therefore, Russ Morash and Jock met with the members to outline our proposal, taking along some preliminary design sketches.

Although commission members are usually well-motivated, their collective decisions can sometimes seem arbitrary and frustrating—as many architects and developers can confirm. To our great relief, dealings with this commission went smoothly. As its members listened to our proposal we could tell at once they found it appropriate. After several meetings, satisfied that the remaining historical fabric would be respected, they gave their unanimous approval to go ahead with the planning. A subcommittee was delegated to examine our working drawings as they were produced, and to deal with whatever problems might arise.

Once the drawings were sufficiently specific about our intentions, they were submitted to the Building Department, and the process of applying for a building permit began. This can be a time-consuming and frustrating experience. For example, permission to change the legal occupancy of the Bigelow House from four to five living units could only be granted by a special session of the Newton City Council, which duly met

and voted in our favor. Later that same evening, it was discovered that the vote had not been valid because there had not been a quorum: it seems that one of the members had gone out at the crucial time to make a telephone call and no one had noticed. This meant a delay of two weeks before the council met again and made the vote official.

To begin with, we were issued a partial permit to cover the demolition of certain internal partitions and the replacement of any rotted wall sills. Next, after assuring itself of the basic stability of our proposal, the department gave us permission to tell them where we put some of the partitions, electrical outlets, etc., at a later date, providing that a complete set of as-built drawings were submitted to them at the end of the job.

We don't mean to discourage anyone, but only to emphasize the importance of allowing plenty of time for the clearance of permits by local authorities. If you get anxious or frustrated at this point, be reassured that most good projects end up being approved.

47

Government grants used to be available for restoring buildings on the National Register, like "Glencairn" (above). With so many historical rehabilitation projects afloat these days, owners must find other sources of support.

Chapter 4

Choosing Priorities

Nothing in a project holds the same excitement as the start of construction. When you first see a place, its possibilities may make you fall madly in love with it, and when you see your fantasies drawn up in architectural plans, you get a sense that the possibilities are real. But when the crew arrives to start work, then for the first time you *know* that your project is actually going to happen. It is also the test of your planning.

At the Bigelow House, as in most rehab jobs, the construction process began with demolition. We knew that we were going to have to rip out all the previous conversion work—the old dormitory rooms and attendant bathrooms—and the architectural plans told us which interior walls would have to come down. As for basic structural repairs, we had been warned that about twenty percent of the house sills (the horizontal wood beams that rest on the foundations and support the house framing) needed attention, so we expected to have to replace some of them. Also, we knew that a lot of the wood shingles on the exterior needed to be replaced. Although our building permit had not yet come through, we had the partial permit to cover demolition, so we were ready to get started.

It didn't seem intimidating. All the new mechanical systems that were to be installed

48

Before you can make something in an old house look the way you want, you have to take down what's there, repair any damage lurking behind, prepare it to take new surfaces, and finally replace the finishes. This colonial kitchen in "Glencairn" was actually the site of an archeological dig as part of its restoration, which provided valuable information about how the original room was built.

would be part of the expense of any new residential building, but here we had a unique site and the romance of an historical building. More than romance, one of the things that excited us most about the Bigelow House as a rehab was the solid construction of the barn and stable buildings. It was inspiring to look at the huge beams and heavy, wide floorboards. I remember jumping up and down on the barn floor to show how solid it was: after all, this floor was designed to support thousands of pounds on the hoof. They just don't build that way any more, and we were sure we had a good investment.

Three months later, however, we were still making structural repairs to the stables and barn. The solid old floor was long gone. We had spent ten weeks jacking up almost 300 linear feet of wall, digging out foundations and pouring new ones. We had replaced ▼ almost eighty percent of the wood sills because they were rotten. Although we had long since received our building permit, the only part of the building where we'd been able to move on from demolition to new construction was the Main House; in the rest of the building, we were bogged down in seemingly endless structural repair. For a while our dream house had taken on

the look of a terminal patient. Norm Abram, the head carpenter, thought that the outbuildings would have collapsed completely in a few more years, and had even gone so far as to comment that it was a good thing they don't build like this any more. The fact that our building had stood for a hundred years by no means guaranteed that it would have survived another twenty-five—without intensive care, it wouldn't have. It is true that the Bigelow House had suffered more neglect than most. Still, it had seemed sound when we bought it, and even with our experience from rehabbing the house in Dorchester, we were not prepared for the amount of demolition and structural repair that turned out to be necessary.

Let's go back to the beginning and see where we misjudged. Demolition work looks straightforward, but there can be many surprises. Like most people, we had made the mistake of estimating the demolition and repair work on the basis of what was visible. Worse, we had trusted the *feeling* of solidity we'd gotten from the building's sound appearance and fine interior finishes. But to have a real idea of what rehabilitation is going to involve, you must be prepared to evaluate the project systematically from every angle, beginning with the structural condition of the building and methodically considering everything that could conceivably have to be done.

Unavoidable Changes and Repairs

First of all, go over a prospective building carefully and make a list of what will have to be done to it regardless of its use. This can include historical restoration if your building is on the National Register or local landmark list. In owning it you may be committed to restoring or preserving it: check at Town Hall and with the local zoning board. In our case, the Bigelow House was a local landmark building, and written into our deed of ownership was the stipulation that we maintain the principal facades as originally built.

Clearly, anything that is rotten or failing will have to be eliminated. That sounds obvious, and in the case of actual structural failure, usually is. In most houses, failure results from the deterioration of materials, although occasionally it happens that a builder has underestimated the loads a structural system must support. Signs of failure are distortions or cracks in walls and ceilings (not to mention outright collapse) and are seldom difficult to interpret.

Deterioration, on the other hand, is often impossible to spot from a superficial inspection, since most major structural elements in

51

Removal of fill between the original porch columns of this building and repair of the cornice and gutter have restored a residential character to this little townhouse in Chicago. Once a neighborhood bar, the Bissell Street "Pump" has been renovated into three condominium apartments.

the average building are covered by exterior and interior surfaces (or in the case of the foundations, by soil), which effectively hide any problem short of failure itself.

It helps to keep in mind that the main cause of deterioration is dampness. In the presence of water, wood rots ▼; masonry freezes, cracks, and crumbles; mortar loses its bond; and insects and fungi thrive. Since moisture enters a house primarily from the ground and the sky, the basement and attic are front lines of defense. Fortunately, in these locations, the basic structure of the house is not usually concealed behind an interior finish, so you may have an opportunity to see for yourself what kind of battle has been fought. Go straight down to the cellar and shine a flashlight on the foundation walls. Modern foundations are made of poured concrete or blocks, but in old buildings you are likely to find fieldstone or even brick. Water, by freezing and thawing, can cause the stones to crack or flake. Water also acts as a solvent for the old lime mortar used before Portland cement was invented, rendering it of little use as a bonding agent in an underground wall. However, dampness on your basement floor, and even moisture seeping through your foundation walls in small quantities need not necessarily pose a serious problem. In fact, many old houses have regular streams running through their

52

basements in the springtime, with no apparent ill effects. You should beware of signs of flooding or inadequate drainage, of course, but since it was practically impossible in earlier times to ensure a perfectly dry cellar, you should be primarily concerned not with water but with signs of damage caused by water. Look carefully for any signs of settling or bulging in the foundation walls, for loose stones or mortar, and for the intrusion of tree roots. Foundation walls support both the building above and the earth pressing in from outside, and if they appear to be yielding at all to either pressure, the consequences can be very serious.

One of the risks we had taken in buying the Bigelow House was to assume that most of the foundations and structural elements that we couldn't see were sound. We could see the condition of the foundation wall in the carriage house where it extended a few feet out of the ground before meeting the shingles, but there was no way to get a look at the subterranean foundations because there was no basement, only a shallow crawl space into which we did not venture. The Main House does have a basement, but we concentrated there on the condition of the wood structure, and took the foundation walls for granted.

The foundations in fact were in very bad condition. They were made of rubblestone,

which is just what it sounds like—roundish stones gathered from rubble or simply picked out of the ground—and much less dependable than the flatter, more uniform fieldstone. ▼ The rubblestone was set in soil that was mostly clay, and therefore very poor at absorbing and draining water. The soil level around the outside of the house was almost even with the top of the foundation walls and wasn't graded to slope away from the house. This meant that rain water or damp earth was constantly permeating even the tops of the foundation walls and the wooden house sills.

In all fairness, without uncovering the framing of the house, we couldn't have identified another important source of the water that damaged the foundations. The house is balloon-framed; that is, the vertical studs are continuous from the ground level right up to the eaves, with horizontal sills for the intervening floors nailed to their inner face. This system leaves continuous vertical cavities between the studs that extend the full height of the house—perfect channels for rain water to be carried down from the roof when the gutters are clogged or downspouts are missing. Since our house had been abandoned for fourteen years, it is certain that the run-off from many rains simply ran in. When gutters clog up, water stands in them and overflows, saturating the wood in

the eaves and fascia boards behind the gutters, and finally the ends of the rafters as well. In our case, the water had escaped downward through the balloon frame, causing the greatest damage at ground level. Not only had water trapped in the soil reduced the rubblestone to plain rubble, but water descending through the walls had caused most of the house sills to rot away as well.

Foundation *repair* is chancy at best. Field-stone foundation walls can be repointed (the mortar between the stones partially chipped out and replaced), or they can be braced, or they can even be supplemented with new walls built outside them. But short of replacing the foundation, there is no easy way to guard against further deterioration, and if you are considering expensive renovation work above, you will not want your investment jeopardized by a continuing risk below. In our case, repair work was out of the question—rubblestone walls are virtually hopeless once they crumble—and since we were determined to make this conversion a quality job, we simply had to build new foundation walls where they were needed.

Another reason we installed new foundation walls was the large proportion of the house sills which had to be replaced. Again, this was something we hadn't anticipated before we got started on the demolition job. From inside the basement, the sills had

looked sound to the inspector—or, at least, most of them had. Once we had bought the house and could start taking it apart, however, we looked behind the shingles on the exterior walls and found that the outer edges of the sills were badly rotted all around the building. In order to replace them, we would have to jack up the house to remove the load from each section as we went along. With the load lifted to give us access to the sills, we could tackle the foundations at the same time.

Foundation replacement, however, is a large job, particularly on the scale we undertook at the Bigelow House. We had almost 300 feet of underground wall to demolish,

Portland cement has too strong a bond to use as mortar in repairing stone walls. To restore this wall, the owner had to chip off a coating of Portland, chisel out the dried mortar which the Portland covered, and repoint the joints with a new limestone and Portland mix.

retrench and rebuild, and each step was a major operation. It's not a job the average homeowner can expect to do himself, since before the foundation walls can be removed the entire structure above must obviously be supported by something else. It is always safer to consult a professional engineer or experienced builder.

Some of the unanticipated extra work at the Bigelow House arose from the need to satisfy certain building code requirements in structures that had not previously housed humans. A good example was found, unfortunately, in those wonderfully solid floors we had admired in the barn. ◄ Even though the floorboards were substantial,

they had been laid on joists only a couple of feet above bare dirt, creating a closed area with no ventilation. This was not common building practice for human shelter in America after the early seventeenth century, and it is no coincidence that few houses built this way have survived. The floor joists were constantly exposed to damp, still air—perfect conditions for rot and insect infestation. We probably should have guessed that trouble was likely; in the Icehouse Unit, the old wood floor had been replaced by a concrete one sometime in this century, ▼ an expense no one would have undertaken unless the old floor had given way. But we couldn't have known for sure about the

condition of the wood below until the floor-boards were lifted up for inspection.

In most old buildings, however, you can examine the framing of the ground floor for signs of insect damage and rot during your inspection tour of the basement. Take an ice pick or penknife with you and jab into any wooden sills, joists, beams and columns that look discolored or mealy, to get some idea of how deeply deterioration has affected them. Pay particular attention to the ends of joists where they meet the house sills or frame into exterior masonry walls—this is a common place for water damage to occur. Look at the bases of any masonry, wood, or steel columns; water that lies there over long periods of time can damage any of these. Above all, wearing your most rugged gardening clothes, do not neglect to visit as much of the crawl spaces as you can stomach. They are usually exceedingly uninviting, but they do reveal the condition of the wood and framing of the house, and you would be making a mistake to buy a house without knowing what stories its crawl spaces have to tell.

As you're making your list of "must-fix-no-matter-whats," take a long hard look at your foundation walls. If they are in poor condition—crumbling, bulging, stones falling out—you can stop right there and walk away. You really shouldn't build your dreams on weak foundations—unless you are, in reality, bolstered by limitless amounts of money. Furthermore, if you are trying to adapt a building for human occupation, keep in mind that outbuildings not intended for habitation were generally built to different standards. This is not to say that barns cannot be grand and durable structures; however, do not forget that they were built for very different purposes than the efficient and comfortable shelter of human beings. Finally, do not neglect to hire a reputable exterminator to supplement your own inspection for insect damage. It pays to be as certain as you can that you're not buying a house with severe insect problems.

So far, I've mentioned some of the basic structural problems and pitfalls to look for in the basements of old buildings. If you're not discouraged by what you see in yours, climb up to the attic and check the framing there. As we said before, gutter maintenance is crucial in protecting the structure from water, and unless it's been diligent, you should look for signs of leaks and saturation at the ends of rafters and joists. Places where the roof is pierced by vent stacks, chimneys and dormers are also potential trouble spots, as are the valleys where two sections of roof intersect. ▶ At the Bigelow House, we were

Rebuilding the third story and adding new window openings, landscaping, and compatible details have rescued an old townhouse in Charlestown, Massachusetts, as an attractive multi-family residence after an interim life as a Knights of Columbus hall.

lucky to have very little trouble with the roof framing, probably because water seepage had been diverted to the basement. The only place where serious damage was discovered was, ironically, the one spot where we had planned a cathedral ceiling. Jock, our architect, had designed a dramatic living room space with exposed rafters for the Woodshed Unit, and we found as we looked closely at the condition of the beams that there was enough damage to make it more economical to take the whole roof down and rebuild it. It's fortunate that our obligation to the Historical Commission did not require us to conserve the original materials because our progress would have been as slow as an archeological dig.

After you've examined your attic and basement for basic structural health, take a walk through the rest of the stories. Inside and out, look for new or newly expanding cracks in the masonry (vertical cracks are not necessarily more menacing than horizontal ones), any crumbling or flaking brick and stone, bulging walls, sagging floors and ceilings, walls and corners noticeably out of square, broken lintels over doors or windows, and gaps between walls and floors. You will have to rely on logic and your knowledge of building connections to diagnose the causes of any slippages you

find. If structural damage is involved, add it to your list of "must-fix." You should also look for water damage within the finished rooms. Rain may have splashed through a broken window long enough to warp or rot the floor. Inspect the areas around the radiator and plumbing pipes as well: many occupants are lazy about fixing a drip.

We were encouraged by what we saw on our first tour of the Bigelow House: everything we looked at inside and out seemed straight and sound, albeit stripped of finery. The north wall of the barn leaned out, and the west wall bulged under the gable, but it was clear that these were due to problems in the exposed foundation walls. Everywhere else the building seemed fundamentally undisturbed, and that is probably why we were confident about the rest of the foundations.

Once you've got a grasp of how much structural decay you're facing, and if you haven't been discouraged, add to your list all the features and finishes requiring conservation or replacement. This includes such things as roofing materials that no longer shed water properly, exterior sheathing that needs repair work to perform its function, broken gutters, rotten window frames, missing panes of glass. At the Bigelow House, the original roof of wood shingles had been replaced long ago by first one layer

58

The original semi-circular and dart-shaped window panes of the Lone Star Brewery proved too expensive to restore when the San Antonio Museum Association renovated the building. Instead they saved the horizontal mullion division and filled in the tops of the old windows with a single piece of glass.

Broken windows, chipping paint, rotting casements and missing slate shingles are typical of roof deterioration on old buildings. The developers of "The Castle" in Newton, Massachusetts, chose the difficult route of restoring all these original details on their building.

of asphalt shingles, then another. The more recent job was still shedding water, but we wanted to replace it for aesthetic reasons. Asphalt just didn't look very attractive on this old house. The roofs are very steep, and are a more important part of the design than in most houses. One of the special characteristics of Shingle Style architecture is the emphasis on continuity of wall and roof surfaces; with new cedar shingles everywhere, the house would look the way Richardson had intended.

A final category of things that might *have* to be changed on your property is site repairs. This includes landscape disasters like a tree fallen across telephone wires, caved-in septic tanks or abandoned wells, impassable driveways, eroding boundaries, etc. If your building is on a piece of subdivided land, you may have to provide new roads and utility approaches. At the Bigelow House there were numbers of things to put on a list of "want-to-change" about the site, but the only thing we thought was dangerous enough to be a "must-fix" was a tree ▼ growing very close to the east wall of the carriage house, whose limbs hung threateningly over the roof. We cut it down. Another "must-fix" was our grading problem around the house, which we would take care of after

the foundations were repaired. The driveway did not have to be moved, but its hardtop surface was full of potholes and would have to be repaired after all the heavy construction equipment had come and gone. Still, we felt lucky to have this site; the pluses far outweighed the minuses. Many fine old specimen trees planted by Dr. Bigelow are still around, and except for a water tower to the northwest, the site is quite secluded. We had a lot of charm without having to do much more than prune.

Now comes the first hard moment of reckoning (assuming you didn't walk away after the first trip to the basement). This is the point to be fully dispassionate about your building, or at least to put a cash value on your emotional response. With your "must-fix" list in hand, telephone various contractors to get estimates for the work. They will generally cooperate as long as you make it clear you're not taking bids, but only trying to plan. If you think you can do some of the work yourself, be honest with yourself about the time and money it will take. Unless you're an old pro at building estimates, when you think you've got a figure for costs, double it; when you've got a figure for time, quadruple it. This is a rule of thumb which Norm's assistant Larry D'Onofrio swears by, and he's done enough rehab work to know how badly a beginner can underestimate. Then, bearing in mind that this expanded budget is what you have to spend before you can even *think* about conversion, decide if you can afford to go ahead with the project. If the answer is yes, move on to the next subject, your architectural fantasies. This is the fun part. Make a list of what you *want* to change.

Things You Want to Change

Your list of what you want to do will contain two kinds of work: things you want to repair or restore, and architectural changes you want to make. If you have an historical building, you might want to remove modifications and restore the building to its original appearance. We did this in our Dorchester house, "This Old House I." On the other hand, you may decide that part of the charm of an old building is the visible record of many different styles and periods of building, but this is a matter of personal taste. If you prefer features of your building's middle age to its youth, by all means enjoy them. Because the Bigelow House had led an interim life as a dormitory building, the upper stories of the main house and servants' wing had been broken up into smaller ▶ rooms. The woodshed had been pressed into service, too, by dividing the space into a

Unmatched doors and closed-up fireplaces are features you may want to change during a rehab. Restoration of this fireplace at the Hellems House in Boulder, Colorado, and removal of modern doors accompanied renovation of the house into five apartments.

central corridor with narrow bedrooms opening to either side. We felt that not only did these modifications obscure the original plan of the house, they were constructed poorly, and useless to our architectural design. In any old building you may find additions or repairs that don't match the original high quality of construction. For your own satisfaction (and perhaps the insurance company's), you may want to bring them up to better standards.

Along with the preliminary architectural changes, you should also make note of all the existing features and fixtures in your building you want to retain, and whether they need special repair work. You may want to conserve ornamental plaster work, carved moldings, stained glass, original light fixtures, and maybe even original bathroom fixtures. Most of these features should be carefully removed before you begin demolition and reconstruction work: they can be stored and refurbished, then reinstated as part of the finish work. The things that can't be moved, like plaster or floors, should be protected by hard coverings. At the very least, tape heavy building paper down over the floors before you begin demolition or construction; this simple procedure costs very little and can prevent a lot of damage.

At the Bigelow House, rehabbing meant both restoration and remodeling. Restoration included the entire exterior of the building, but on the inside was limited to the Main House. Most of the banisters, wood mantles and paneling, lights and bathroom fixtures, and even copper pipes had been ripped off by scavengers. The wood trim—baseboards, door and window frames—and most of the original paneled doors in the main house and servants' wing would come off and be taken to a wood-stripping place, where old paint would be removed and the pieces returned to us for installation and refinishing. We found other "finish" material we wanted to save in the other structures, such as the floors in the barn. The interior walls of the carriage house and horse stalls were finished in matchboard, or cut tongue-and-groove boards, ▶ which had a nice look after one hundred years, and we thought we might be able to re-use it somewhere or possibly sell it to someone who was looking for old fir paneling. Generally, though, the conversion of the Bigelow stables and barn to modern apartments involved new interior construction and remodeling of existing spaces. Our work really fell into two categories: restoration of the Main House, and adaptive remodeling of everything else.

Remodeling plans will depend on the kind of re-use you have in mind. Converting a

If you are fortunate enough to find architectural details as beautiful as this mosaic floor in "The Castle", be sure to protect them with temporary coverings before you begin demolition and renovation work around them.

This unusually decorated Ionic column at the McCosh House in Princeton, New Jersey, was held together with electrical tape until restoration contractors could duplicate the carving.

A new balcony and double-height living room were created by cutting out part of the second floor in this converted coach house in Andover, Massachusetts. A steel tie-rod replaces the missing joists as bracing for the vertical side wall.

large, single-family residence into a multi-family dwelling requires functional decisions about which areas will be public and private, and provision for equal and adequate access to services like garbage collection, mail distribution, fire-escape routes, laundries, storage, and so on. Some of this is governed by local housing codes. If the building you're converting has never been residential, it is usually a challenge to plan for all the necessary circulation routes and living patterns.

Unusual and unique spaces are one of the most appealing results of converting old buildings to new residential use. You frequently have the chance to work with much larger floor-to-floor dimensions than ever possible in modern residential buildings, so that future apartments can be designed with high ceilings, loft spaces and visual complexities that could never be achieved in a high-rise. One of the easiest places to gain a sense of architectural space and its possibilities is in an old barn, where the volume is often the same as in a typical colonial house, but the structural framing is exposed and the space is continuous. The hard part is deciding what kinds of spaces you'd enjoy most. The basic opportunities in re-using space are to open up existing spaces to make larger ones or to break up larger spaces into smaller ones, horizontally or vertically. There are many possibilities, but before you set your heart on something, keep in mind its structural requirements.

Taking out interior partitions completely is only possible if they are not load-bearing; bearing walls can be removed, but the weight they carry must then be supported by columns or a deep transfer beam.

Cutting away a piece of ceiling to create a double-height space is a popular architectural move, but it can only be done easily in the direction of the floor joists; you can remove entire joists, but you can't saw off sections of joists without adding a new bearing wall or columns under the free ends. In our Barn Unit, Jock was able to open up the living room ceiling under the west gable because the joists were running north-south. He planned the opening to parallel the direction of the joists, not the floorboards.

Removing ceiling joists to expose an attic for a cathedral ceiling may take away the structural ties between the walls of a room, as was the case in our Woodshed Unit. If we had simply removed the joists, the weight of the roof would have pushed out the walls, and the roof would have collapsed. One way of preventing this is to replace the joists with a couple of exposed steel cables or tie-rods. Old architecture is frequently reinforced by

64

The living room of this private residence in Eastern Long Island, New York, shows very clearly the spatial characteristics of the original barn. The owners and architects chose to leave the structure open for a large, airy living space.

The steel framing of a former candy factory in Philadelphia, Pennsylvania, was exposed when the architect decided to open up the space vertically. The triple-story height creates a dramatic, sky lit stairwell and dining area in this new apartment.

these modern ropes. In the Woodshed, we rejected that approach for aesthetic reasons; the rods might not be very noticeable in a real cathedral, but this room was not going to be very large. We suspected, too, that the Historical Commission would cast a dim eye on such a bionic procedure. We finally decided to take the roof down and reframe it with a heavier ridge-pole and deeper rafters, tying the pole plates together around the perimeter as a substitute for the horizontal ties removed with the ceiling joists.

Other structural limitations to your design are inherent in different kinds of construction. It's easy to add a window opening in a wood-framed wall ▼ because most of the weight of the wall is carried by its main framing, and the new window does not have much to support. When you make a new puncture in a masonry wall, however, the window framing must bear the weight of the stone above the opening. Since this can be extremely heavy, the new opening must be framed in with steel headers, or stone must be removed in order to leave an arch that will transfer weight around the void. This can be a considerable operation because you must support the wall until the arch or header is in place.

The existing structure of your old building

will determine whether you can re-use it the way you've planned. The basic structure, including the window locations, should not have to be changed very much. If you're hoping to subdivide a building for new residential use, measure the windows to make sure you can lay out practical living quarters that conform to code without changing the pattern of fenestration. Column spacing and bearing wall locations will also affect your layouts. Most existing structures, if they're sound, will meet load requirements for residential use, but if, for example, you want to add a new mezzanine floor within the original floor-to-floor height, you must make sure the structure can support the addition.

Besides setting limits to your aesthetic fantasies, the existing structure of a building can make a difference in practical planning decisions. It is important to check whether the original structural framing of the building will accommodate openings for stairways, elevators, exits, and other circulation changes and safety measures your plans require. Bathrooms and kitchens have to be located where water can be piped ◄ to and from them. Bathtubs full of water are heavy items, so you must locate them where existing structure can support them. Drain pipes and air ducts you might intend to install as part of new mechanical systems may not fit within the existing dimensions of an old wall.

All of this is really the beginning of another list. After you've decided what kinds of architectural and landscape changes you'd like to make, itemize separately all the things you'll have to do to effect these changes. If this sounds like a boring amount of paper work to do before you ever really get started on a project, keep in mind that it's the only way to reduce the financial risk you're taking. Every renovation job will contain some surprises, but careful homework at the planning end will keep those to a minimum.

Things You Must Change

The next step is to make decisions about the new utility and energy supplies and equipment—where they will go, and how you will accommodate them. This list should also include the changes you'll have to make in the building to satisfy code requirements for the kind of re-use you want. Be *sure* to do research on the local situation before you sign a contract to buy any property or you may find yourself with a building you can't re-use and can't re-sell.

Most American cities and towns are controlled by zoning ordinances and, sometimes, subdivision ordinances as well,

67

When the Customs House in Boston, Massachusetts, was converted to offices and apartments, attic space was reclaimed for living, and the wooden structure of the roof became the main architectural feature of the rooms. Exposed plumbing pipes are almost unnoticeable next to the huge scale of the transfer beams.

When re-using space in old commercial buildings for apartment living, architectural planning must work around existing column spacing and window openings. In this large living room in Printers' Row in Chicago, the fireplace chimney has been inserted in a wide space between two small windows of a side elevation.

which have specifically to do with land use and the environmental impact of building. Your first question about a re-use project should be whether it is possible under current zoning or whether you can get a variance to make it possible. If you can reassure yourself on land use, you must next make sure your plan can satisfy local requirements for sewage disposal, fire department access, parking, building size, and sometimes even considerations of visual homogeneity. Such requirements will be found in the zoning ordinances and in local building codes. Take the time to visit the Building Department in your city or county to find out which codes are used. Sometimes you can buy copies of the codes at Town Hall or from the publisher. The codes will determine what kind of building you can erect on the kind of soil and sub-soil you own, so if you're hoping to do additional building, be sure to check out the geological characteristics of your site. The building code will also regulate (according to fire-ratings) the kind of construction and materials you may use in your district and will limit the size and occupancy of any building from the standpoint of human safety, rather than environmental effect and property value.

Most cities and all states have their own building codes, which are modified versions of one of three national codes—Basic Building Code, Uniform Building Code or National Building Code. Each national code was drawn up by a different professional group and is used predominantly in one part of the country or another, but there is no way of predicting which you'll find in your state. Fire codes, health (or sanitary) codes and housing codes may exist in addition to or be incorporated into the local building code. Usually the only difference it makes is in who enforces them. However, you should acquaint yourself with all the regulations in whatever codes apply to your property and be sure you can comply. It is sometimes possible to effect changes in code by legal action, but this always takes time and money. You must decide how important it is to you, and don't believe anything you hear about bribing building inspectors.

The electrical profession is the only building trade with a uniform, national code, and it will vary from state to state only in small local modifications, such as whether an owner is allowed to install new wiring himself. No matter who installs it, new electrical work must be inspected by a licensed engineer, or the building inspector, for compliance with the national code. Plumbers have a national code, but it is a minimum standard that you will find in effect only in places

An extreme solution to land-use or zoning problems is to move the building in question. This maneuver is difficult but possible even for quite large buildings. The McCosh House in Princeton, New Jersey, has been moved at least twice; the latest move was only several hundred feet but involved just as much work as a longer

one. The detail above shows some of the technique used in moving the brick Robeson House in New Bedford, Massachusetts, which was supported on steel beams and then rolled along a wooden track which was built for it.

where local codes don't exist in great detail. Plumbing ordinances are usually written into local building and sanitary codes. Your building will have to adjust to all of these codes and ordinances, no matter how many years have passed since it was built and how few code requirements its original builders faced. Decide what improvement you legally must make, and add them to your list.

You should next make a schedule of demolition work and new construction necessary to make the actual plan changes. For us, for example, this would have included all the dormitory partitions as well as the back stairs in the servants' wing, the wall between the Woodshed and the Icehouse ▼, the old dividing wall in the Barn, the stairs to the hayloft, and the roof in the Woodshed. Repair and new construction can be estimated in a general way unless you have a speculative set of architectural plans showing exactly how much new material you'll need.

Things You Can't Change.
While you're listing the things you want to change and have to change, don't forget that there is a category of things you *can't* change. Preservation requirements may forbid you to change the exterior and certain

interior parts of your building. We couldn't alter the original design of the principal facades at the Bigelow House.

The actual structure of the building is something you can change very little. List which walls and chimneys are load-bearing or somehow critical to the integrity of the structure. On a more general level, list the zoning and code requirements that will affect your plan layout and that can't be changed. Decide where on the property the municipal services (water, sewage, electricity, telephone) are supplied. You can usually move those utilities around on your own property but you will still have to link up with the city in the same place. Finally, you should consider what site limitations there are and list any major and unmovable obstacle to your plans.

A Program and a Price

Now you're in a position to make up a tentative program for demolition work, repair work, site work, and new construction. Correlate the lists of things you have to change, things you want to change, and things you can't change. Some of the entries will cancel each other, leaving you with a pretty good idea of what you can expect to accomplish. In order to evaluate the costs of this program, you could rely on bids from

general contractors, but it is a good idea to try to estimate the time and materials yourself. This demands a certain amount of research into contractors' manuals. They will tell you how to estimate the amount of a given material you'll need for each operation and how to figure its cost. When Norm makes up a bid for a job, he uses a guide that gives him a logical way to walk through the building with the architectural plans and itemize the materials he'll need. But be methodical and cover everything.

As you work your way around the building, you may decide that the time and materials balance sheet favors rebuilding over restoring in some places. We wasted a fair amount of time and labor at "This Old House I" trying to patch up existing conditions in some places, only to find later on that we had to rip them out anyway. So, we resolved not to save old construction just because of its age.

When you have an idea of what this adaptive use will cost, you must decide if it is feasible for you. You can go over your tentative program and eliminate some costs, or replan the scale of operations entirely to come up with a smaller figure. When you review your program, you may realize that you can't reduce your conversion plans and still expect to recover enough of the expenses

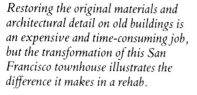

Restoring the original materials and architectural detail on old buildings is an expensive and time-consuming job, but the transformation of this San Francisco townhouse illustrates the difference it makes in a rehab.

when you sell. We didn't take this approach to our project, and it's hard to say whether we would have gone ahead with the Bigelow House if we had. Since we were planning to film the renovation for television, we probably would have done it anyway, even knowing the worst of what we would be getting into. The site was convenient, the house was both photogenic and of historical interest, and we had a lot of local support for the project.

Possible Scheduling Problems

One last essential element of planning is to try to foresee what kinds of snags may occur in realizing your program. For example, you will need to know how soon construction can begin. This will depend on several factors—how soon your financing will be available, how soon an acceptable contractor will be free to start the job, and how soon you can get demolition and building permits. Financing is the most obvious starting point, and also tends to be the least ambiguous factor. Once they have the money, many people assume that contractors are sitting around just waiting to be hired. Some certainly will be, but you might consider that the contractors you would be most interested in would be the least likely to be waiting for work. You may well be lucky

and find someone good who does have time, or if your job is big enough, you may find good contractors willing to make time for it, but as a general rule, keep in mind that you will probably have to wait for someone good. The same thing goes for the sub-contractors too, and although you won't need most of them until later, you should start lining them up as soon as you have worked out a realistic schedule with your contractor.

Above all, don't neglect one important detail—the building permit. You will be very lucky if it is only a detail! In most localities, the law requires you to obtain a building permit before beginning any construction (in some places you even need a permit to cut down a tree). Either you or your architect must submit a set of plans, including plumbing and electrical work, to the local Building Department with an application for a building permit. Sometimes this is a routine procedure; in large cities it can be a slow and infuriating bureaucratic process. If your time schedule is tight, you can usually get a demolition permit to allow you to begin stripping away old plaster and fixtures while you're waiting for the city to approve your application. This is how we started.

The application for a building permit can be quite complex. If you are changing the

71

This late 19th century Connecticut church has, in the late 20th century, been transformed into a spacious private home. In converting the nave to living-dining room, the high vaulted ceiling was painted sky blue and is visible through open beams which were added to define a more personal living scale beneath. The kitchen

illustrates, with its lovely stained glass windows, the merit of retaining some of the original features of an old building.

use of the building, you will have to account for occupancy changes and all architectural, structural, and mechanical changes governed by code. You will have to prove often by listing building regulations right on the plans, that there are no building violations. In addition to the architectural plans, your application will usually have to include structural details of alterations or new construction, plumbing diagrams, electrical plans, and a topographical map or city land plan showing the plot you own, and the building's location relative to the surrounding properties. It is sometimes difficult to find out from the Building Department all the things you will be required to submit, in how many duplicates, and in what order. You may want to hire an architectural consultant to expedite your passage through the labyrinth, especially if you're working in a large city with a mammoth bureaucracy.

These *expeditors* are professional architects or engineers who are familiar with all the requirements of the local Building Department. In places like New York City, they are indispensable. They can inform you, much more efficiently than anyone in the Building Department, about all the applicable regulations and how to cover them in your permit application. If the application requires the signature of an architect or engi-

neer, and you don't have one already, you would be wise to engage the services of such an expeditor to sponsor your application. You can find one by inquiring at the Building Department (if he's any good at his job, he'll be old friends with the application reviewers), or by asking other architects.

Once you get a fairly clear idea of when work can begin, you must start thinking about how long it is likely to take, and what kind of constraints weather or the availability of materials may impose. This kind of scheduling is one of the things your contractor should do for you, but again, you should also put some thought into it yourself. Try to anticipate any possible delays, and figure out what kind of financial repercussions they might have.

Having covered everything you can think of, you may be prepared for as much as half of what you will actually encounter. Every old house has its idiosyncrasies. *Expect* to be surprised—you can see that we were at the Bigelow House, in spite of having already completed "This Old House I." Nevertheless, if you plan your construction strategy thoroughly along the lines I have described, you'll be ten times more prepared than most rehabbers.

72

When this schoolhouse in Charlestown, Massachusetts, was converted to apartments, the attic space with its large wooden trusses and round ornamental window provided a more interesting living room than most new apartment houses can offer.

Construction Planning

1. List what *must* be changed:
   Rotten or failing elements in basic structure, finishes and features, and site
   Historic preservation requirements.

1A. Given what *must* be done, can you still afford to consider the project?

2. List what you *want* to change:
   Remove anachronisms, shoddy construction.
   Repair features you want to save.
   Suit the building to your new use by installing new services, systems, and architectural features.

3. List what you will *have* to change in order to achieve the re-use you would like.
   Changes required by Code:
   zoning, subdivision, fire protection, building safety, health (plumbing), and electrical energy.
   New equipment and systems: purchase, installation.
   Additional construction necessitated by your plans: demolition, strengthening existing structure, new structure, and finishing work.

4. List what you *can't* change because of: zoning, subdivision requirements, preservation requirements, basic structure, municipal services, utility supplies, and site limitations.

5. Correlate lists 1 through 4 and produce a tentative program for: demolition, repair work, new construction, equipment and installation, and site work.

6. Evaluate costs. If necessary, adjust lists where possible and re-evaluate costs. Repeat this process until you have developed a feasible construction program.

7. Develop a tentative schedule, considering:
   How soon can you get financing?
   How soon can you obtain a building permit?
   How soon can a good contractor begin work on the job?
   How long will the job take, given the size of the crew your contractor will probably use?
   Are any necessary equipment or materials in short supply in your region?
   How hard will it be to schedule subcontractors when you need them?
   How much will delay cost you per day?

73

The landscape architects who use this old water mill in New Jersey as their offices have made a dramatic entrance out of the old machinery room over the millrace. Offices and conference rooms are located in the former storage areas.

Chapter 5

Designing the Spaces

From now on, we thought of the Bigelow House not so much as a single building, but as it had been originally—a well-knit group of units, each with its own special purpose. In the five new living units, one of our main objectives was to emphasize the individual character of each. Even the original names were retained: it was much easier to say "the Icehouse" than "Living Unit Number Three."

In this attempt to diversify, our approach differed from the average developer, who must rely on standardization to keep costs down. Our reason for wanting five different types of kitchen, for example, instead of one standardized unit, was to demonstrate to our viewers as many different approaches as possible. Similar problems do not have just one solution, and there is no one perfect solution to any layout.

Many of our choices were the result of compromises and tradeoffs: the kitchen in the Main House, for instance, might have been as large as the original one, but for the fact that this would have taken too much space from the kitchen in the Woodshed Unit, against which it had to be backed in order to share a plumbing vent stack for the sake of economy. Similarly, the party wall at the other end of the Woodshed Unit could not be pushed back because of the desirability

Several different kinds of old, shingled structures were moved to a new site and linked together to form a single residence and guest house in this Long Island re-use project. In concept, the result is not unlike the Bigelow House, but because these barns and house were never designed to exist under one roof, the effect is completely different.

of letting the living room of the Icehouse Unit extend to the south wall for better views and lighting.

The Main House Unit

The Main House was the only unit that had been entirely residential. Because of this— and the fact that it had been well planned to begin with—the amount of re-design required was relatively small. It could be described as the *restoration* unit, as quite a lot of the original interior detail had been saved and stored for re-use.

As might be expected, this was the biggest and tallest of the units, consisting of three ▼ floors and a basement. On the ground floor is an entrance hall with an elegant staircase— by far the most elaborately detailed item in the whole building—a spacious living room looking southward down over the meadow, a small toilet and, in the ell, a generous dining room with an open kitchen. On the second floor is a large master bedroom (the same size as the living room), with a bathroom, a recessed washer and dryer, and a smaller bedroom with its own bathroom. The third floor contains two bedrooms with a shared bathroom and a long storage space going back into the roof of the ell. The principal features consist of the ornate staircase in the

hall, six fireplaces with most of the original tiles and trim either restored or duplicated, the sunbath opening off the living room, the turret in the corner of the master bedroom, and the belvedere on the roof.

Even in the early stage of planning it seemed clear that this unit, the largest, would also be the most formal. Many of the principal features and the more sophisticated details were found here.

For instance, the living room and kitchen had a unique type of window ◀ divided into three single-pane sashes, which slid upwards out of sight into a large hollow pocket in the outside wall. In summertime the windows simply disappeared and were replaced by screen doors that literally "came out of the woodwork" from concealed pockets in the side of the openings, thus opening up the interior of the house to a wooden deck, flush with the grass, that extended most of the way around the main house and ell. It is possible that Dr. Bigelow himself thought of this idea for health reasons, in order to have the best possible ventilation in the heat of the summer.

Whether or not this was the case, however, it anticipated by about forty years one of the principal tenets of the Modern Movement in architecture—the continuity of interior and exterior space. In the old photo-graphs that were discovered later, these triple-hung windows had bottom sashes divided into two panes instead of one. But Jock decided that the slightly simpler versions that he had already ordered were not significantly different, and so no change was made. (Because they take a long time to be delivered, windows are one of the first items a wise builder will order.) Similarly, the hipped roof he had designed for the sunbath was allowed to remain, although the old photographs clearly showed some form of lean-to roof; the hipped roof appeared in several of the sketches made in Richardson's office and we all felt that it fit in well with the rest of the design.

In the Main House, the principal interior spaces remained more or less as originally designed and with the same functions. The only real modification was the addition of a compact kitchen adjacent to the dining room in a space previously subdivided into pantries and a passage to the old kitchen (which was now the new kitchen in the Woodshed Unit). Jock saw the efficiency kitchen as ideal for people who might not want to spend long hours cooking but who would occasionally entertain formally. In order to open up the space, however, it was decided not to have complete separation between the cooking and eating areas so only a

77

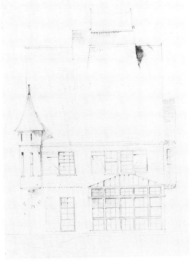

The original sketch from Richardson's office of the south elevation of the Bigelow House shows a hipped roof on the sunbath and no roof at all under the turret, but the rest of the architectural features and the building's proportions are there.

Another sketch of the front (east) elevation presents roughly the same building configuration as was built, but because the shape of the windows is different, the house looks more like a cottage. This is one of the few sketches which does not include the gable room over the front door, only a conical roof which is drawn to look like part of the shingle skin stretched out.

1
First Floor Plan

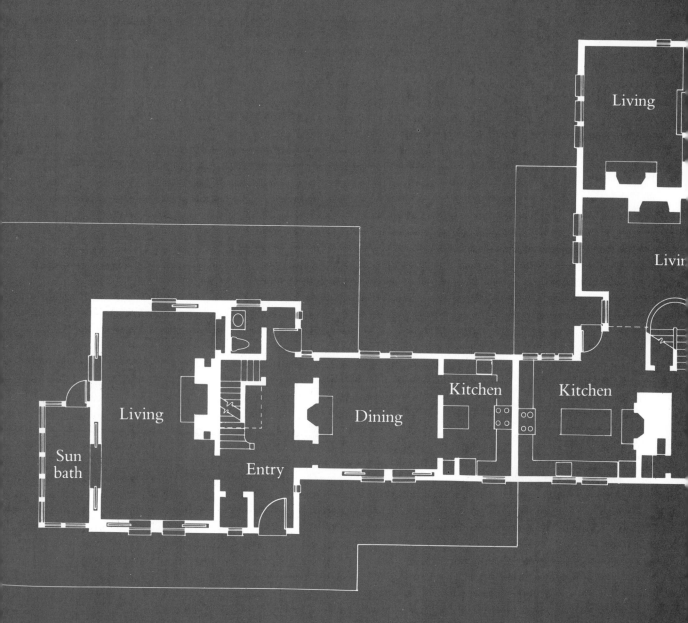

Living

Livin

Living

Kitchen

Kitchen

Living

Dining

Sun
bath

Entry

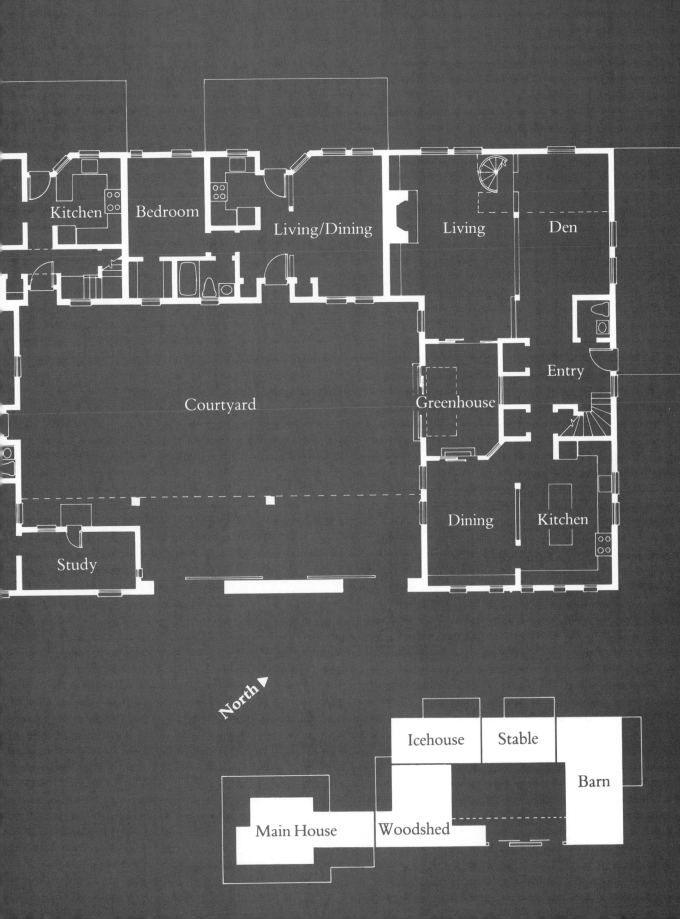

Kitchen

Bedroom

Living/Dining

Living

Den

Courtyard

Greenhouse

Entry

Study

Dining

Kitchen

North ▶

Icehouse

Stable

Barn

Main House

Woodshed

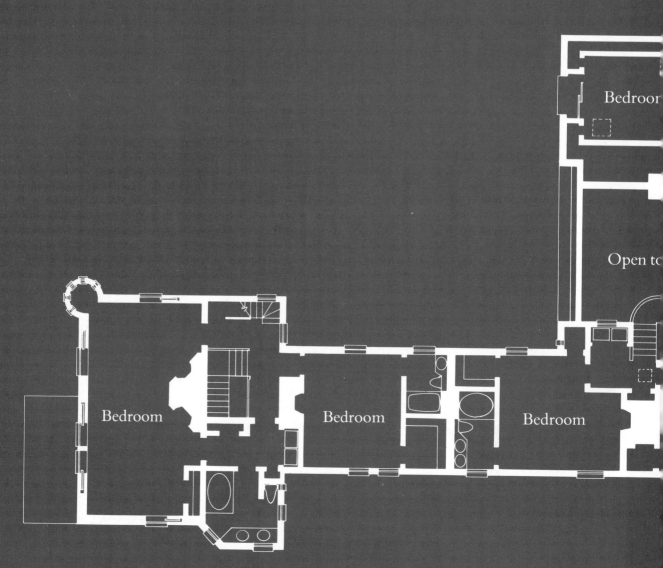

Bedroor

Open to

Bedroom

Bedroom

Bedroom

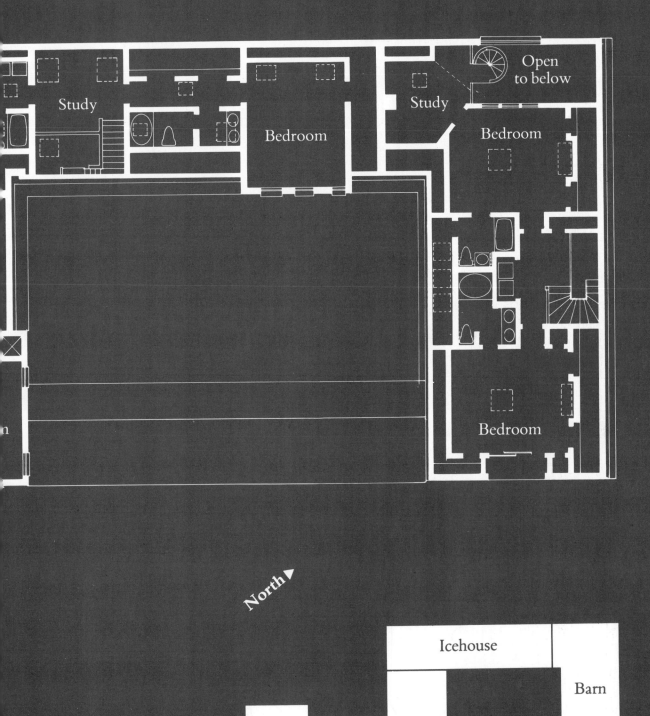

Study

Bedroom

Study

Open
to below

Bedroom

Bedroom

North▶

Icehouse

Barn

Main House

Woodshed

3
Third Floor Plan

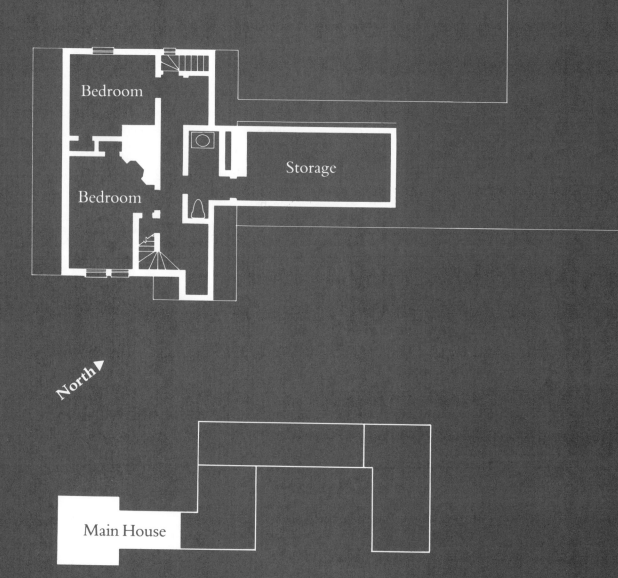

Bedroom

Bedroom

Storage

North▶

Main House

low wall was used between them. Because of the highly efficient through-the-range air extractor, there would be no problem with cooking odors in the dining area. The kitchen counters and cabinets were laid out along the other three walls, with a small island against the low wall. The sink was centered on the one window with a full view. The other window, facing the more public side of the building, has a sill five feet off the floor— one of the original windows which, at the request of the Historical Commission, we agreed not to enlarge.

The dining room is slightly reduced from its former size to give a little extra area to the kitchen, but it's still quite spacious. ▶ The original fireplace remains, on the wall adjacent to the hall. On the east wall are two of the triple-hung disappearing windows that open out onto the wooden deck, convenient for serving meals there in the summertime. (It is ideally situated for sunny summer breakfasts.) On the west wall of the dining room are two windows of the same width, but which do not extend down to the floor. There was some doubt that these windows had four-over-four sashes to match the rest of the house, but it seemed illogical to mix two types of sash in one room so we opted for the larger single panes.

In some of the design sketches from Rich-

ardson's office, the main house was turned at an angle of almost thirty degrees to the rest of the complex—probably so that the living room would face exactly south. (What we have been calling "south" in the text for convenience is actually southeast.) The final decision to make the living room face southwest does not seem to have been a bad one, however, as plenty of morning sunlight from the east can enter through the large triple-hung window on the side wall. Although not enormous (it is about thirty by fifteen feet), the living room seems spacious. This has much to do with its major feature— the sunbath—which looks straight down 83

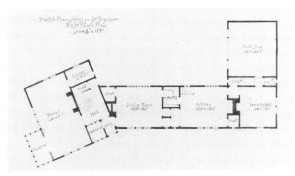

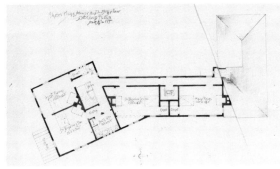

An earlier plan for the house, drawn here without the stable court and barn, turned the formal living areas of the Main House so that the rooms were oriented to the south.

the hill over the sloping meadow and across the countryside. The view is further dramatized by the trees on each side of the meadow. The living room is also graced by four of the large disappearing windows and an almost monumental wood-paneled fireplace. ▼ To one side of the fireplace, a small recessed bar closet has been built with a sink that shares plumbing with a new toilet behind the ornamental staircase in the hall, an area formerly used by Dr. Bigelow as a small study.

The Main House had two features that were extremely useful for energy conservation. The extra thickness of the three outer walls of the living room and the master bedroom above it (necessary for housing the disappearing windows and the sliding screen doors) provided space for extra thermal insulation, a necessary compensation for the exposed hilltop location. These walls can be seen clearly in the plans. The winter heating bills will also be reduced considerably by the sunbath that opens off the living room and provides what is now known as *passive* solar heating. On a sunny winter day the sunbath will be quite warm, so that simply opening the glass doors between them will raise the temperature in the living room. A small low-energy electric fan could be used to distribute the warm air more effectively.

On the second floor, the large master bedroom was formed by merging two smaller rooms, each of which had a corner fireplace skewed at forty-five degrees and placed back-to-back, and these were left as a sort of sculptural unit projecting into the new ▶ space. As this still implied some division of the space, the thought was that it might be used as a study/bedroom. The turret in one corner seemed to have been designed more for external effect than internal practicality, being only about four feet in diameter. The slit windows, too, with all their romance, proved to be not so effective for viewing the landscape. Nevertheless, it remains a piece of harmless fun—a "folly" as it would have been called in the nineteenth century. The original windows had been double-hung, but we decided to replace them with fixed sashes; the system of counterweights was quite complicated, and the windows were not necessary for cross-ventilating the room.

An especially interesting space on the second floor is the bathroom, which juts out from the main gable of the house to form a gable of its own—another feature in many recently designed houses. This can be seen on all the early sketches, including the one reputed to be by Richardson himself, so it is definitely an intrinsic part of the original design.

85

Nevertheless, when the walls were opened up, there was evidence that the design had been changed during the construction. Also, the whimsical forty-five-degree cut-away window (a feature most often seen in Queen Anne-style houses) may have been introduced as an afterthought: an unused connection for a rain leader had been installed where the two gutters above the window intersect —right in the middle of the present window. Jock felt that the room had probably been originally intended to be rectangular, with a small slit window in that corner on one of the walls, but that at some time—either during or after construction ◄—the forty-five degree window was introduced. Anyway, the cut-away corner window does give a better view when using the plumbing, and as an added touch of luxury we added a large half-sunk whirlpool tub.

On the third floor, the two small bedrooms and the bathroom remained as before. A surprise feature, however, was the addition of a long storage loft in the roof space of the ell. We felt this space was too useful to waste, although it was not high or wide enough to make useful living accommodation and it was difficult to reach. This was solved by placing a panel door in the bathroom. The area then provided just the sort of storage for larger objects so often over-looked by architects and clients.

Fortunately, most of the original details of the main house had been saved—largely due to the efforts of John Howard—and, after a visit to the paint stripper, could be returned to their original places. Missing parts were carefully duplicated and then blended in with the original. In addition to the panels on the living-room fireplace there were baseboards, moldings, and fireplace, door and window trim. There were also enough of the original tiles still left in several of the fireplaces so they could be re-set and intermixed with new duplicates. One of our luckiest finds was a sample of each of the three types of spiral baluster from each step of the big ornamental staircase. When we first saw the house there was nothing left of the balustrade but the newel posts at the top and bottom of the stair, and we almost gave up any hope of trying to restore it. Now, however, with the balustrade samples, we had almost enough evidence to reconstruct it perfectly. The one missing link was the handrail, and so we did the next best thing by using a handrail from another Richardson house as a model for this one.

While discussing the problems of the staircase we discovered that we had a nearby source of highly skilled craftsmanship. Richard Husher, the husband of Elsie

Ready-made banisters and balusters are available in historical styles, but if you want to restore a missing piece or duplicate an original, you will have to resort to custom carpentry.

Husher (who had played a pivotal role in saving the Bigelow House from destruction), proved to be an expert wood lathe operator —something he does as a hobby. He turned out beautiful replicas of the balusters for us.

We were also fortunate to have the original panel doors in the house. ▲ Like many other items, they merely needed their numerous coats of old paint stripped before refinishing. In an older house doors can often be quite a problem to replace. The ideal solution, of course, is to duplicate them if there is enough money to do this kind of custom work. But sometimes an alternative must be found.

88

Some stock copies of old doors really don't look very authentic. On the other hand, plain flush doors can look rather incongruous surrounded by an elaborately molded frame. But like so many of the decisions in this sort of rehabilitation, it's up to you to judge your taste and care for detail up against your pocketbook.

The Woodshed Unit

Although most of this unit is actually part of the ell of the main house, ▶ the largest room (the living room) occupies the site of the former woodshed. As the woodshed was the only part of the entire structure that had to be totally rebuilt because of rot, this unit is unique in the project because it combines restoration (doors and trim in everything but the living room) with totally new construction.

On the ground floor, in addition to the large cathedral-ceilinged living room, there is also a large kitchen, a study, and a small toilet. Upstairs is a bedroom with a large walk-in closet and an adjoining bathroom with a large, half-sunk tub, and a second bedroom with its own somewhat smaller toilet and shower. There is a washer and dryer located near the bedrooms in an alcove off the upper hall.

The design of the kitchen sets the style for the whole unit. There are basically four

Where privacy is not an issue, one way to bring more light into an old building without sacrificing the convenience of existing doors is to replace the door panels with glass. This sliding door between kitchen and hall allows the cook to isolate cooking odors without blocking off light from the back of the house.

ways of laying out a kitchen: the single straight counter, "Pullman" or parallel counters, the L-shaped counter, and the U-shaped counter. All sorts of variations are possible, of course. The first two options seem to be used mainly where space is restricted or where there is either an entrance at one end or two entrances.

The Woodshed Unit kitchen is actually the biggest of all the kitchens at the Bigelow House. It is also the only L-shaped plan, which is particularly suitable for kitchens in which people can sit at a table and eat. The work counters are confined to two walls, so there is less of a sense of being surrounded by cooking equipment. In the Woodshed kitchen, the feeling of openness is further extended because there are no doors to the living room. Both rooms appear to flow into one another and thus the kitchen becomes one of the family rooms.

The contrast between this kitchen and the Main House kitchen is very significant and shows something of the wide range of possibilities in the remodeling or adaptation of older houses. Kitchens make a good illustration, because they are the most complicated and personal rooms in the house, but the same range of possibilities also applies to the other rooms.

In Newton, as in most cities and towns, only two dwelling units are permitted between fire division, so we had to build a good, solid wall between the Woodshed Unit and the Icehouse Unit. We incorporated two back-to-back fireplaces there—each in the center of a living room wall. In the living room of the Woodshed Unit the centralized fireplace adds the final touch to what is an almost completely symmetrical room. The fireplace is echoed on the opposite wall by a projecting stair landing that gives a very effective balcony-like view of the room and by the recessed front and rear entrances. A high-pitched wood cathedral ceiling further emphasizes the symmetry, as does the window placement, which also ensures good natural cross-ventilation from the southwest breezes of summer.

Another item of particular interest is the small study that projects out under the roofed arcade. This has been particularly well-placed for maximum sound insulation, being separated from the living room and the kitchen—the main centers of noise—by the dining area.

The Icehouse Unit

This unit is located in the narrowest wing of the whole complex, limiting the possibilities for room layout. Here the main rooms and

the circulation and serving spaces had to be accommodated within the width of the building, which is about eighteen feet. On the second floor, the usable space was even narrower due to the low, pitched roof. ▶ Therefore, the spaces in the Icehouse Unit are generally smaller than those of the others, and, on the upper level, tend to be elongated.

This unit, in addition to occupying the area of the former icehouse, also extends on the upper level over the top of the Stable Unit and includes the former groom's room as one of the bedrooms. On the ground floor, the living room, dining room, kitchen, and toilet are compactly arranged; on the upper floor there is a spacious master bedroom with its own bathroom/toilet and generous closet space, and a second bedroom with adjoining bathroom. At the head of the stairs there is a large open study area through which the bedrooms are reached. A washer and a dryer are located in an adjoining alcove that leads to the second bedroom.

Washers and dryers are found in all the units except the small Stable Unit, and much thought was given to their placement. The Main House was the only unit that allowed the option of the traditional location—the basement—but it was felt to be too far away from where clothes were actually stored. Al-

Planning kitchens in small, reclaimed space is often a challenge in re-using old buildings. Each of these kitchens, one in a barn and one in a renovated house, is extremely compact, but the effect is not at all claustrophobic because the kitchen area is well-planned.

ways thinking about plumbing economy, we also took care to place them so that they would not need a separate stack.

Again, the kitchen reflects the character of the Icehouse Unit—compact efficiency. Here it is fully separated from the dining area by a full wall with a doorway. The dining area, however, is merged with the living room in order to open up what is really a relatively small area. We also lowered the floor in the living room, which is particularly effective when coming in from the relatively narrow and low-ceilinged corridor.

The fact that all this is accomplished within an area of about sixteen feet by forty feet demonstates how effectively space can be used. It also shows that good adaptive design is much more than a two-dimensional exercise of pushing a few partitions around on a flat sheet of paper!

Because the original icehouse obviously had no windows, they had to be added— again, mostly away from the public side of the building. To take full advantage of the magnificent view down over the meadow and the south exposure, three new windows were grouped together on the south wall of the living room, matching the four-over-fours of the main house. Also, in the kitchen, a new recessed door/window unit with the window raked at forty-five degrees helped

open up the view down the hill from that room. On the outside wall above the new living room, what had been a large loft door became a window for the second bedroom. In order to retain as much of the feeling of the original opening as possible, however, the new window was recessed several feet back from the outside wall face, forming a small balcony for plants—a device Jock was to use again in the Barn Unit.

The former groom's room was the only area in the entire unit that had been used for human habitation, so it had windows, which were restored. It was also the only area on the second floor with full headroom across

92

This auto body shop in San Francisco had enough light from the large, overhead skylight to permit architects to re-use the space as an office without needing windows in the side walls. A central corridor planted with trees and ferns gives the same impression as an exterior view.

the entire width of the building. ▼ As a little more light was needed, however, two skylights were inserted unobtrusively at the other end of the room in the west slope of the pitched roof. Similar skylights were used to light the other spaces on the upper floor—all of which were inside the roof space—causing far less disruption to the original design than the alternative of creating new dormer windows.

About half way along the ridge of the roof over this wing and somewhere in the vicinity of the groom's room there was a picturesque ventilator cupola, of the type seen in old carriage houses. ◄ As well as ventilating the

stables, it casually echoed the shape of the romantic corner turret of the main house. No real use could be found for it in the new layout, but it was nevertheless retained and restored as a significant feature of the original design.

The Stable Unit
This is by far the smallest and most compact of all the units, consisting simply of a living/dining room, bedroom, kitchen, and bathroom all on one floor. At one time we had seriously considered making it one large area for living, eating and sleeping in order to achieve a feeling of space, but this idea was

abandoned, since we felt most people would want at least a separate sleeping area. The kitchen—although it is only about one-quarter the size of the family kitchen in the Woodshed Unit—is superbly organized and provides highly efficient cooking facilities.

In order to keep the price of the unit as low as possible, some corners were cut: the finishes and trim, for instance, were more standard than in the other units and there was no fireplace—although a *thimble* connection to a new flue was provided for the addition of a wood stove. This new flue was part of the second firewall required by the building code, and in it we embedded the new fireplace for the living room in the adjacent Barn Unit.

Features of the Stable Unit include a recessed rear door similar to the one in the Icehouse Unit, but this time in the living/dining room: the skewed window again providing a wide view in combination with the two straight windows alongside. There is also an unexpected bonus in the form of a very useful walk-in closet in the bedroom, which provides ten feet of space.

Adequate storage area for clothes, coats, etc., is often overlooked in the heat of decisions about the big spaces, and the unfortunate owner often must improvise something later. In older houses, closets are often quite

small—people used to have fewer clothes, and wardrobes (or free-standing closets) were often part of the bedroom furniture. The shallow depth of the closets is puzzling until we realize that the clothes were usually hung flat against the rear wall from a hook, instead of at right angles to the wall on a rod. To convert these closets to the two-foot-clear depth that we now consider minimal is often difficult, and you may have to rework the general layout to some extent in order to introduce closets that really do their job properly.

One rule when planning closets is to try to build them into the walls instead of sticking out into an otherwise regular space. This is usually possible, but it may require an experienced designer. Sometimes a bank of closets can be built in, either side-by-side or back-to-back, as a partition between rooms, making an effective sound barrier as well as a place to store clothes.

Other forms of storage should not be forgotten. A closet near the main entrance for coats, hats, sports equipment, and so on is an essential. Also, a place must be found to store larger objects—bicycles, extra furniture, etc. It should not be necessary to give up prime living space for this, but some reasonably dry place in a basement or attic will do for things that are not often used. Just be

Doors and cabinets built into the wall of an attic staircase in this house take advantage of the depth of the space under the stairs for added storage.

sure the basement is not subject to severe dampness or flooding and that the attic is accessible—as in the Main House Unit. Our other units did not have basements or attics and so the problem of general storage was solved by incorporating it into the new garage building, making it the garage/storage/solar collector building.

The Barn Unit

This unit, the second largest, has the most contemporary interior and shows the most imaginative use of space. During construction, the entire unit had to be practically rebuilt because of excessive deterioration, and

this gave us a freer hand in our design than might otherwise have been possible. On the ground floor all the main spaces flow into one another, and parts of the upper level have been opened up to continue the visual connection vertically. In the living room the two levels are connected by an elegant spiral stairway.

The ground floor consists of a small entrance hall partly open to the upper level, a spacious living room and study partly open to the upper level, a small toilet, and a formal dining room half-open to the large adjacent kitchen. For privacy, the living and dining rooms have comparatively small windows on the courtyard side, but between these rooms is a large greenhouse built into the building with large window/doors facing into the courtyard. This greenhouse is accessible to both rooms as well as to the courtyard by sliding glass doors. A dominant feature of the living room is the fireplace built into the new party wall—a spectacular piece of rough stone masonry.

By placing the main entrance to this unit on the north side and limiting its connection with the courtyard to the greenhouse, Jock provided an extra degree of privacy. The greenhouse, combined with three skylights, forms a light well which brightens the second floor. All the skylights, which were

The architectural possibilities of re-using a large barn include simple open planning with balcony rooms, or elaborate spatial gymnastics with lots of added structure and color.

needed to light the top floor, had to be carefully placed to avoid spoiling the visual effect of the eyebrow windows.

The entrance hall has two unusually attractive features: a glimpse of the greenhouse through the wall facing us as we enter, and overhead natural lighting. Surprisingly, this comes from one of the skylights high up on the south slope of the roof.

On the upper floor is a master bedroom and a second bedroom, each with its own bathroom. Off the second bedroom is a small study that is also accessible from the living room by means of the spiral staircase. The master bedroom contained the old loft door opening; a recessed window was installed in it. The new window in the other loft door opening in the west gable—the least public side of the building—was not recessed, as it opened into the new two-story space with the spiral staircase. To give extra light, another window opening was made directly above it in the apex of the gable, but was kept separate to preserve the original outline of the loft doors.

Because of the new two-story space over the living room, the second bedroom did not extend to the outer wall; it was lit, therefore, by a combination of skylights and an internal window that borrowed light across the open well from the window on the outer wall. One of the bathrooms is lit by a similar device, borrowing light from the light well above the greenhouse by means of a wheel window. The other bathroom takes advantage of one of the eyebrow dormers ◄ which gives a surprsing amount of light for its size.

One of the most distinctive external features of the Barn Unit was a row of five tiny windows, nearly square, placed high up on the ground floor of the east gable. These were carefully preserved and incorporated in the new layout to give clerestory lighting to the dining area and kitchen. The window at the south end was slightly different in proportion to the other four and divided into two panes instead of one. To date, no one has been able to come up with a rational explanation for this mystery—perhaps there is none!

104 Windows

During the first stages of the design phase, with no reliable as-built drawings or photographs to go by, all our decisions had to be made solely on the evidence of the existing fabric or by conjecture. This was especially true of the windows. Fortunately, John Howard had done a scrupulous job of saving and storing all the remaining sashes before the windows were boarded up; when these were examined, many of them proved to be 99

Replacement of a 19th century window in this coach house was achieved by filling in the area above a modern, ready-made sash with wood siding. Perhaps because of the room height inside, the fill has been added above rather than under the sash, so that the window appears to have slipped down below the upper story

as old as the house. Although we knew up to this point that they would be the standard double-hung type, there was no evidence of how they were subdivided. Jock had drawn them as having six subdivisions in each sash (commonly known as a *six-over-six*), although a more probable late nineteenth-century configuration might have been the six-over-one seen in most Queen Anne and many Shingle Style houses. It came as quite a surprise, therefore, to find that most of the original sashes for the house were of a quite unusual four-over-four pattern, thus producing a rather casual assortment of horizontal rectangles as subdivisions. The visual effect was less restful and consistent than the

100

more traditional vertical rectangle would have been, but tended to reinforce the somewhat wayward image of the rest of the building. To what extent these window subdivisions were a deliberate part of the design is unclear, however, since it seemed unlikely that Richardson himself would have supervised such details.

Despite our successful research, the design of the replacement windows still posed some problems. There was now no doubt about what the originals looked like and all that had to be done, it seemed, was to duplicate them and everyone would be happy. Things seldom go this smoothly in the real world of construction, however: we needed to replace traditional single-glazed sashes with double-glazed sashes for energy conservation. But in order to do this we would have to use heavier *mullions* (the narrow wooden bars that subdivide the sashes), thus changing the appearance of the windows considerably.

The decision to use double-glazing had not been made lightly; double windows would have done much the same job, but it seemed to us that single sashes with air-sealed *sandwiches* of glass would be easier to operate and would also resemble the original windows more closely. In order to house the relatively thick sandwich panes, however, more wood would have to be milled from the mullions, weakening them to the danger point; they would then have to be widened in order to be strong enough to withstand normal stresses. The original mullions had been the traditional three-quarter inch width, but the window manufacturer now recommended that this be increased to one and one-quarter inches.

In many parts of a building an extra thickening of a half-inch might pass unnoticed, but in the case of mullions this is emphatically not so. Being highly visible against the window, they become one of the most visually sensitive points in the whole design. This visual weight and the profile section of these little sticks plays a role out of all proportion to their actual size in determining the basic character of a building. If the windows are the eyes of a building, the mullions must surely be the pupils.

Generally, a well-designed mullion is deeper than it is wide—about twice as deep —to give it maximum strength in a fore-and-aft direction against wind and other pressures while, at the same time, allowing the maximum amount of daylight to pass and reflect from its sides. Narrow mullions tend to look light and graceful, whereas heavy, wide mullions can look like prison bars.

The Newton Historical Commission balked at the proposed wide mullions because they would have a considerable effect on the appearance of the house. At first we felt a certain righteous indignation because the decision had been made in the interests of saving energy. But we had to admit that they had a point. Jock then went back to the window manufacturer to see if there were any alternatives and finally managed to get the width reduced to seven-eighths of an inch. This compromise satisfied all parties.

The story of the windows has been told in some detail because it typifies problems that arise when historical restoration is combined with adaptive use. Such problems, of course, do not arise in pure restoration work, which aims at recreating what the building used to be. Nor do they exist at the other end of the scale where the appearance of the existing building is not considered a major factor.

On the subject of window replacement, here are some useful *dos* and *don'ts*.

If the original sash design is unknown or cannot be copied, make sure the design of the new one is the same period and style as the house, *e.g.*, six-over-six for Federal and Greek Revival houses, two-over-two for Italianate and Mansard Style houses, and six-over-one for Queen Anne houses.

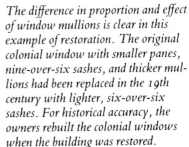

The difference in proportion and effect of window mullions is clear in this example of restoration. The original colonial window with smaller panes, nine-over-six sashes, and thicker mullions had been replaced in the 19th century with lighter, six-over-six sashes. For historical accuracy, the owners rebuilt the colonial windows when the building was restored.

Never try to age a house by using sash divisions of a period earlier than that of the house itself—don't put Colonial twelve-over-twelves in a typical nineteenth-century house.

Make sure that the new window units fit the old apertures. Nothing looks worse than bits of paneling put up to fill the gaps between the original aperture and some standard window that doesn't fit it properly. Custom windows do not cost much more than standard sizes and are well worth the money.

Do not introduce shutters or external blinds unless the house originally had them, or was intended to have them. (One exception to this might be a house of no great historical significance that has large bare expanses of wall.)

If shutters or external blinds are being added, make sure they look as if they would work even if they don't: they must each be the full height of the window and at least half its width.

Shutters and external blinds are one of the most effective ways of preventing the sun's rays from penetrating the window and heating the interior; think seriously about making them work to help cool the house in the summer.

102   It seems that most of the external design problems in adapting any house have to do with windows and doors. Of doors, all that can be said here is that it is best to introduce as few new ones as possible, and to keep those few off the principal faces of the building. In the Bigelow House, all the new doors were kept to the west and north sides. Among other details that may have to be added are new chimneys, which should match the existing ones. We built two new stacks ▼ for the use of the Woodshed, Icehouse, Stable and Barn Units, using brick and mortar similar to that used in the Main House. Great care must be taken to match not only the bricks but also the mortar joints

—both in color and thickness. Tying-in with old brickwork can be done successfully only by using old bricks, as new ones are slightly less high and so must have wider mortar joints between them to line up with the old courses. If you patch an old wall with new bricks, particularly when bricking-up windows and doors no longer used, the difference in thickness of the mortar joints will show up as a permanent and very visible ghost.

General Design Principles

Here are some general design principles that apply to the outside and the inside of build-ings being adapted for new uses.

Don't try to force inappropriate or unworkable transformations on a building; try to find out what the building might *want* to become, and go along with it as far as possible. The Bigelow House seemed to be made for subdivision because of its original diversity of use.

Make as few alterations as possible to the original fabric of the building; this usually makes good sense both aesthetically and economically.

Avoid irreversible changes; remember that today's aesthetic meat may be tomorrow's poison, and give any building of merit at least a sporting chance of being restored to its original shape some day.

Never try to age a building by adding details from a period before its time. This is architectural forgery, and it is never convincing.

Try to keep all necessary changes to the outside of the building limited to the least public side. This applies to such things as fire escapes, new dormers, window and door alterations.

Try to incorporate many of the major features of the original design in the new plan—the principal entrances, staircases, fireplaces, paneling, and molding—and treat them as assets.

103

Chapter 6

Demolition and Repair

After the working drawings are done, your architect will submit the plans to local contractors for competitive bids. It is usually his job to find out who the good contractors are in your area, but you can do it yourself. We didn't seek bids on the Bigelow project because we had Norm Abram lined up from the start. He had worked with Jock Gifford, our architect, on similar projects, and, of course, he was responsible for the success of the Dorchester house. With Norm to run the job, we could act as our own contractor.

Finding a Contractor

If you are soliciting bids on your own, and aren't familiar with the local contractors, ask around to get recommendations from people who have recently rehabbed a building. You may find that some of the best small firms don't even advertise because they don't need to. Call up a number of contractors and arrange to go over the plans and the site. It doesn't hurt to get as many bids as possible —don't make a decision without at least three figures. What you would always hope to find is a perfectionist young contractor who needs to establish a reputation and who is therefore willing to offer a low bid in order to get work.

If you don't find such a person, you will have to decide among a scale of bids that will

"Restoration" is not a cut and dried procedure. In some projects, historians and conservators work hard to make new repair work invisible, but this often confuses archeologists and students of a later generation. Another approach is to duplicate workmanship and material without hiding repairs.

The restorers of this thatched, half-timber cottage in England, for example, have allowed their modern repairs to become a distinguishable part of the building's aesthetic.

reflect the various contractors' approaches to business. You may be surprised at how much these can vary, which is why it is worthwhile to get a number of them. Some contractors will price themselves very high because they don't expect to be competitive, but prefer to rely on the small percentage of clients who will pay for a certain level of quality. Needless to say, they don't get many jobs, but the ones they do get are invariably profitable. Among the more competitive bids, some will still be fairly high— you must decide whether it represents a good job or a good profit. Others will bid low in hopes of getting the contract, and may then cut corners in the construction or charge you for lots of extras to make up their profit. Don't, as a general rule, automatically jump at the lowest bid: you'll probably pay for it. Another temptation is to hire the person you get along with best. A pleasant personality is obviously no indication of his ability as a builder. Ask him for references and follow them up. Look over his other jobs. Talk to previous clients as well as to plumbers, electricians and other sub-contractors who have worked with him. How he gets along with his co-workers as well as their opinion of his abilities are very important factors.

A good protection is to write as many

specifications into the contract as possible. They can be included in the detailed architectural drawings making up the bid set, or can take the form of written specifications for materials, products and building methods. The more detailed you can be, the more accurate a bid the contractor can make. You should also establish the terms of payment in the contract. Because of the unknowns involved, many builders will feel more comfortable signing a rehab contract on a time-and-materials basis, which means calculating your bill using an hourly labor rate and the cost of materials purchased. The danger, of course, is that this kind of arrangement provides incentive for a contractor to drag his feet at your expense. On a small job with a contractor you trust, the incentive isn't large enough to matter, but you should avoid undertaking a big job on a time-and-materials basis. Flexibility can be worked into a normal bid contract by adding a provision leaving you the option to switch over to time-and-materials whenever it seems desirable, such as in the case of cost overruns, while still ensuring that it is in the contractor's interest to work as efficiently as possible.

On the legal side, you must be certain that any contractor you hire carries liability and compensation insurance. Ask him for the name of his insurance company and write to

it for a Certificate of Insurance. This is normally included in the process of obtaining a Building Permit, but if for some reason it isn't, don't neglect to get hold of the certificate. Without it, you can be held liable for your contractor's workman's compensation payments and any injury claims that might be made with relation to the job.

However, if you do feel at all uncertain about making these kinds of assessments, you can hire an architect or engineer to monitor the progress of the work. One of the distinct advantages of hiring a good architect in the first place is that he or she is responsible for overseeing the quality of construction. Occasionally, other contractors on the project will tell you if someone else is doing a bad job. I have had a builder warn me that the man installing the burglar alarm system was violating both electrical and fire codes, and I have had a plumber tell me that the best electrician was not the man I was about to hire but someone twenty miles away whom I hadn't considered.

The Construction Sequence

One morning, after weeks and months of searching, negotiating, and planning, trucks pull up in front of your empty old house; the crew has arrived. For the first time in perhaps fifty or a hundred years, the sound of

The Tulley House in Sante Fe, New Mexico, is an adobe house which was painted to look as if it were made of brick when a fashion for expensive eastern materials reached the Southwest. Charmed by this architectural anecdote, restorers were careful to reproduce the painted "brick" courses when they renovated the building.

hammers echoes through the house, bringing it to life.

Where does the process start? Norm generally follows the same set sequence of demolition and construction for virtually any rehab job. If you are handling the administration of your project yourself, you'll have to be familiar with the order. In a large job, you may well find yourself at different stages in different places, but for each location, the order of business is the same:

- Remove all finish work, trim and features you want to save.▼
- Gut the interior where planned.

- Do structural work
  —new, additional structure
  —repairs to existing rough carpentry or other basic structure.
- Re-do your roof, if necessary.
- Build framing for new partitions, non-structural features.▼
- Do all your sub-contracting work (plumbing, electrical, heating and air conditioning, insulation etc.).
- Have the work inspected by the Building Department.
- Do all the finish work (plastering, cabinetwork, flooring, hardware, etc.).

Salvage

Demolition is literally the construction process in reverse; you start with the previous builder's finish work. The first thing Norm and his crew did, after covering the floors, was to go around the main house and the servants' wing taking down the doors and wood trim that we wanted to save. Unhinging doors is simple, but removing woodwork is meticulous work, far from the usual image of demolition as great free-wheeling fun. To remove wood frames without splitting them, Norm uses a small crow-bar with a flat edge he slips behind the wood to work it loose, so that it can be pried off in one piece. Once it is free, he removes any nails (from behind, so the surface of the wood isn't dented), marks it for later identification, and stores it. Another handy tool for this kind of work is a *nail-pull*. Since it can pull out nails from the front of a piece of wood with minimal damage to the surface, it can be used when you don't have room to insert a flat bar or where you're particularly afraid of splitting the wood.

There was so much vandalism at the Bigelow House that there weren't many features left to protect or conserve. Four-fifths of the building we were converting, after all, had never been finished inside the way a house would have been. Even in the Main House,

108

there was almost nothing to save of the original decoration and finishes except the wood framing around the doors and windows. Everything of value that could be removed was either thrown out when the building was remodeled for dormitories or stolen during the years it stood empty. There were no light fixtures left. There were a couple of old bathtubs and sinks, but no really nice hardware or fixtures in the bathrooms or kitchen. Anything made of valuable metal was gone—all the copper plumbing pipes, even the copper flashings from the roof. As for wood detail, all the banisters and stair rails were missing from the main staircase,

leaving one solitary newel post. Some of the original wood paneling remained around the fireplaces in the living room and the master bedroom upstairs, and we carefully removed it for repair and to duplicate missing pieces. Actually, the house had never been treated to the kind of ornamentation common in residences of the wealthy at that period. It was intended to be a simple summer cottage, and so a minimum of expense was devoted to sumptuous materials and decoration. At some later date the living room walls had been paneled in vertical wood boards,▼ but they were anything but sumptuous and not at all in keeping with the

original design. We were eager to pull them out as soon as possible.

Demolition

Once the decorative, or finish, materials we wanted to preserve had been removed or protected, the next step was to gut the interior of the house of all the finishes that had to be replaced and any partition walls we planned to eliminate. The next step was the exterior finishes, and this was a very large job. For one thing, we had all the shingles on the exterior walls and several roofs to remove. In the interiors, we planned to tear down all the partitions put up for dormitory use in the Main House and the Woodshed. In addition, there were the original plaster and matchboard finishes we had to take down to install new plumbing, wiring and insulation, there were stairs in wrong places and there were a few original walls to remove.▼

Our first step was to get rid of all the plaster and the dormitory partitions. To avoid confusion, we had spraypainted big red X's on all the condemned walls. ▶ The outside demolition crew Norm had hired then gamely put on dust masks and attacked the old finishes with crow-bars and hammers. They didn't remove the framing of any original interior walls, however, until all the ceiling joists and stud walls had been exposed, so that we could be sure of the engineering consequences of any planned change.

Before you begin this kind of demolition, make sure that the water and electricity are turned off so that you won't cut into a wire or a water main and be electrocuted or flooded. We knew that these services had

110

been turned off in the Bigelow House for years, so we only had to pull out the old systems as we uncovered them.

Old bathtubs, toilets and sinks were carted away. None of ours was worth saving, but don't assume that no one will want an old tub or sink. Lots of rehabbers are eager to locate antique fixtures, and some contractors are willing to buy castaways for their stockpiles. The plaster ceilings and wall surfaces came down in great heaps of broken lath, dangling wires, mouse nests and plaster dust. It's very satisfying to make this kind of mess, and especially so here because someone else was cleaning it up, separating the

wood lath out of the rubble and shoveling plaster out the window into our dumpster. Without the plaster we could see through walls and for the first time we began to sense what the new rooms would be like. ▼ The Woodshed area was transformed into one large space; the upstairs study and bedroom seemed more than twice as big once you could see the volume as continuous.

There are interesting optical illusions during the construction process, and it takes experience to learn how to visualize at each stage how the spaces will seem when finished. A floor plan, even with furniture drawn in, often gives a sense of grander di-

mensions than the built spaces. However, when you first see the floor areas defined only by low foundation walls, the rooms may suddenly seem alarmingly small. Then, when the walls are framed in with studs, the rooms seem, paradoxically, to grow, only to shrink back a little as finishes and fixtures are applied. Finally, to your relief, the scale will grow again when furniture is brought in and arranged. We were a long way from couches and chairs, however, as we tore down the old finishes and watched the spaces grow before us for the first time.

While the plaster crew was working on the Main House, Norm and several workers began taking apart the inside of the barn and the horse stalls in the Stable. The wood stairs up to the hayloft were removed, as well as the matchboard wall that separated the carriage berth from the stalls. This wasn't a bearing wall, so it could be removed very easily. Now we could see the entire ground floor area of what would be the Barn Unit—almost 1800 square feet of open space, most of which would remain open in the new scheme. When I say ground floor, I mean it literally. ◄ Norm had taken up the heavy planking of the old barn floor to examine the condition of the subflooring, and we discovered that there was nothing below the floor joists but dirt. This meant that even though

the floor joists were big heavy beams, without the added support of a subflooring they had been weakened by constant exposure to moisture and insects. We were going to have to provide greater insulation and protection for the floor of a family's living and dining room, so the original floor had to come up completely to allow for further excavation and preparation. We were careful to store the boards so that they could be re-used.

Structural Demolition and Repair

We had begun demolition during the third week in March, and it took the plaster crew only about two weeks to clear the interior of the Main House, servants' wing and Woodshed. Norm used this time to start checking out the structural damage around the house that we couldn't have seen before. He was a bit suspicious after our discovery under the barn floor, so he went outside to look at the condition of gutters, eaves, shingles and wood trim. ▼ We began tearing off some of the shingles at foundation level to get a look at the wooden house sills. The first place we looked was along the south and west walls of the Main House, and the sills turned out to be in terrible shape. From inside the basement they had looked sound, but on the outside face the sills were badly rotted, leaving very little good wood to support the weight

of three stories. This discovery marked the
beginning of long and laborious demolition
and repair. In the end, we replaced nearly all
the sills and most of the foundations all
around the perimeter of the entire complex.

At one corner of the Main House, the sills
were level with the soil and sloped in such a
way that they had been in steady contact
with ground moisture as well as rain water,
and were rotting. ▼ It looked as if the dampness had attracted insects, too, though we
didn't see anything crawling around at the
moment. Ideally, a foundation wall should
lift the wood structure of a building above
the ground level by one or two feet in order

to prevent this from happening. Shingle siding is also designed to divert water from the sills: at the base of the building it sweeps out like a skirt so that rain water will be diverted away from the base of the house (assuming that the earth is properly graded to permit the water to run off). As we dug away the soil from around the foundations, however, and pried off the lower rows of shingles, we learned how thoroughly the water had defeated our old house's defenses. It was discouraging to discover just how much unforeseen repair and replacement we had ahead of us.▼

At the same time, more cheerful reports

came from the mason we called in to examine our chimneys, and from the exterminator who came to inspect our old house for pests. The mason had had to put up scaffolding on the roof to reach the chimneys of the main house because the pitch of the roof is too steep to climb. I felt safer in the cherry-picker the cameramen were using, but since I was supposed to appear in *front* of the camera, I made my way up hand over foot. It was soon clear to me why most homeowners don't (and shouldn't) attempt roof work by themselves. After the depressing news about the sills, I was prepared to learn that all the bricks were well on their way back to dust and that the entire chimney would have to come down. I thought that maybe expecting the worst would keep me from losing my balance when the verdict came. Once bricks in a chimney or side wall begin to absorb water, freezing weather quickly causes them to crack and eventually pulverize. As they deteriorate, moisture leaches through them into the house, causing problems wherever it ends up. Luckily, this had not happened to us. According to our expert, the bricks needed repointing, but they were still structurally sound and reasonably water-resistant.

Our mason suggested that after repointing the joints we should paint the brick with a

silicon sealant to increase its water resistance. This is a modern miracle coating which supposedly makes masonry surfaces watertight. Some of these products can cause severe trouble if applied in the wrong place, as they can trap moisture *inside* a brick wall, and thus actually increase frost damage. Furthermore, even if the glaze on your bricks is quite old, the primary route of water penetration is through the joints, which means that your first concern should be with keeping them in good repair. Because mortar won't bond properly to a silicon-coated surface, a sealant (which will have to be renewed every year) should not be applied until the joints have been carefully repaired. Before going ahead with any new finish products, you might want to consult a building conservationist, whose advice about preserving materials in your climate will be reasonably scientific and unbiased.

Not only were our chimneys in good shape structurally, but we even found that the fireplaces were still usable! Our mason examined the firebrick linings and found them sound in every location. Of course, each chimney needed a good cleaning to remove rubble, birds' nests and accumulated creosote, but chimneys should always be cleaned once a year.

Our first idea had been to install wood-burning stoves in all the old fireplace openings, but we abandoned that plan for several reasons. In order to provide effective heating, stoves should be free-standing rather than pushed back into a fireplace. A free-standing stove must be installed at least a minimum distance from any wall, and even so, wall surfaces nearby must be protected from heat by a fire-resistant material like ceramic tile or asbestos-board. It was clear that if we installed the stoves efficiently, we ▼ would lose the old fireplaces with their mantles, wood paneling, and original tiles. This hardly seemed a worthwhile sacrifice. A real-estate consultant confirmed our feeling by pointing out that wood-burning fireplaces are still one of the most desirable features of a house or apartment, and that it would be unwise to cover them over. In the end, the only wood-stoves we installed were one in the Woodshed Unit, where it did not replace an existing fireplace, and one in the Stable Unit; otherwise, we left the question of stove installation up to future owners.

Insect Damage

Another encouraging report came from the exterminator. He showed us where there had been some powder-post beetle work ▶ (behind the gutters and in some of the sills—

all places that had suffered some water damage) and explained what to look for in an insect inspection. Termites can't be seen, as they actually work inside the wood, but where they enter a building they make little mud tunnels you can look for on your basement walls. Powder-post beetles can be spotted by the piles of sawdust and excrement they drop where they work. They attack the outer layers of wood, leaving it mealy and soft with visible channels and holes, which will be filled with dust. Carpenter ants also chew up wood, but instead of digesting it like the beetles, they move it into piles a few feet away from where they are working. Rather than living in your house, ants are likely to be commuting from a nearby nest, and if you can locate the nest, it is usually easy to destroy. Beetles and termites require more extensive fumigation techniques in and around your house, and especially where termites are concerned, it is best to rely on a professional exterminator.

Subterranean termites are the only house-damaging insect that can be affected by preventive measures. Generally, sound construction that minimizes dampness in your structure is the best discouragement to other bugs. This means that the very process of rehabbing was our best protection against the insect world, since the inspector didn't

find any evidence of active infestation. We turned our attention back to the job of repair.

Structural Work Begins

An easy place to start the structural demolition was with the little sunbath on the south side of the Main House. At this point, what had once been glass was now just plywood sheets. Even though the roof eaves had nice details, the structure of the porch and its floor were quite rotten, and we had known from the beginning that we would have to rebuild it entirely. Until we received old family photos of the house, we hadn't even known that an open corner porch had stood under the turret, so we had made no plans to restore it. The sunbath didn't take very long to come down. We knocked out the plywood boards, then the posts holding up the roof, and finally wrenched off the roof itself, leaving an unsightly horizontal gap in the siding of the house. ▼ The whole operation took less than a morning. Another day we took away the old iron fire-escape. Without that and the boarded-up sun porch, this old house was regaining a little of its dignity.

Foundation and Sill Work

Once the Main House was gutted, we could go ahead with the major structural job on the sills. Norm's examination of the shingles and eaves had revealed extensive rot all over the west wall of the servants' wing, so we decided to start there and work our way around the Main House. Our first plan was to jack up the whole side of the house to remove the weight from the sills while we laid in a new *plate* (the carpenters' term for the horizontal sill plank). ▶ Norm got out his house jacks and set them up so that they would lift the floor, and we pumped. Nothing moved but the floor. The three stories of wall didn't budge. We couldn't understand what had gone wrong until we realized that we had special problems because we were dealing with a balloon-framed house and not with platform framing. The walls were framed *beside* the floors and did not rest on top of them. In order to lift the house we had to design a second framing system that would tie the floor and the walls together so they would move as one piece. All in all, it took several of us three days of thinking and preparation to devise a way to raise the wall off the foundation a couple of inches. Once the wall was free, we could cut out the original sill and fit in a replacement. When the whole sill isn't rotten and you only need to

replace a segment, it is easiest to cut out the bad part and drop in a new piece, but generally we were faced with entire lengths to substitute.

The foundation replacement was a slow process. First Norm had to remove the old rubblestone, which was a tricky job with a backhoe since there wasn't much room to maneuver. ▼Then he dug a deeper trench for the concrete footing, or base pad, which anchors the foundation walls to the ground below frost level. Steel formwork for the footings had to be set up before the concrete could be poured. We rented the formwork from a concrete company. It comes in modular sizes that can be adapted to almost any foundation dimension, although sometimes we had to fill in corners with wood forms. As the concrete went in, we vibrated it mechanically to help distribute it evenly and inserted horizontal steel reinforcement bars along its length. Then it had to cure for several days before the formwork could be removed and set up again for the foundation wall on top of the footing. ▼Footings here are continuous for the length of the wall, and Norm made them about eighteen inches wide by eighteen inches deep. ▶The foundation walls were either eight or twelve inches wide, and centered on the footing. The process of pouring them was similar to pouring

122

the footings, but we had to be more careful to level the top surface so that the sill would lie straight, and we added a series of street tie-bolts that we could connect to the sills later to prevent them from slipping laterally. After the formwork was removed, we put down a creosote pad to act as a moisture-barrier between the concrete and the wood sill and as an expansion joint between the two materials. ◄ Before Norm filled in the earth around the new foundations, we painted them with a tar-based coating to shed water on the side next to the earth. ◄

Once we had begun the sill and foundation-wall replacement, the process seemed never-ending. We had to excavate and replace foundations everywhere around the complex except on the east facade, where they were still in good shape. We felt lucky to be able to save the old stone wall there, since it is the principal facade, the one anyone sees first when driving up to the house, and the stone foundation wall was an important visual element of the elevation design. Because of the scope of the sill and foundation work, we had to replan our construction schedule. We decided to concentrate on the Main House, so that at least one unit could proceed at a normal pace while we caught up with the large structural repairs in the other units.

124

Scheduling the Construction

Norm still followed the logical sequence of operations, even though he was working at a different stage in any one unit at any given time. For the next six weeks, at least four different stages of construction were in process simultaneously around the courtyard. Only once did we find it necessary to deviate from the normal sequence. In the Main House we began the sub-contracting and finish work before the new roof was put on. Ordinarily we would have finished the roof first, so there would be no danger of water damage to our new plaster, wiring or insulation, but since the roofers weren't too far behind us, we took temporary measures to protect the interior, and went ahead.

After the end of April, then, the demolition and construction process took on the look of a circus. The Main House was definitely center ring. By the beginning of May, we had taken down all of the original partitions that were to be removed there and in the Woodshed, and we were beginning to frame up the new ones.▼ Subcontractors began arriving. The plumber had come to rough in the water systems for domestic water, sewage removal and radiator pipes. Wiring, piping and insulation were installed, new closets were framed, the new windows

began to go in, bathtubs arrived, tile was laid, the boiler was installed and the pipes hooked up. The new framing and wiring began in some areas of the Main House even while Norm was still working on the sills and foundations of the other units.

Before moving on to the plastering and finish carpentry, therefore, we had to wait until all the structural repair was complete and we could make sure the framing was as square as we could make it. In addition to the sill work, Norm and the carpenters replaced the rotten sections of fascia under the eaves and sections of exterior sheathing that had deteriorated under the water-logged shingles. ▶ When all was ready, we checked the alignment and found that the whole house leaned a little to the east. ▼ Norm was able to push it gently back into square with the backhoe, and we secured the new joints. In many old wood-framed buildings, fallen out of line because of structural failure or deterioration, it is impossible to straighten out the deflection in old timbers, so we were lucky. In spite of the state of the sills, there had been little shifting in the Main House; we had to jack up one floor that had been pulled out of square when a chimney settled, but basically it was a very straight old house.

As the interior work took over in the Main House, demolition and structural repair

went on in the other units. The outdoor crew was taking off the old shingle siding so that repairs could be made to the sheathing, and roofers were stripping off the double layer of asphalt shingles on the roofs (strictly speaking, this was aesthetic rather than structural repair). ▶ Once the foundations and sills had been finished at the Main House, Norm devoted his time and carpentry skills to the large structural jobs in the Barn, Woodshed and Icehouse Units. Virtually all the foundations around all the units had to be dug out and replaced, which involved the same amount of bracing, excavating, formwork and concrete pouring as at the Main House,

126

multiplied by a factor of three or four.

In the Barn Unit, the entire north wall had been ruined by water damage and foundation failure, and was scheduled to be rebuilt. Because it was such a long wall, holding up half of the huge barn roof, it was a big job to brace up all the weight involved and take away not only the foundations but half the wall as well. ▶This kind of radical surgery occurred all around the courtyard. We lost the walls and roof of the Woodshed Unit, the north and south walls of the Barn, and the east and west walls of the Icehouse.▼

It was disconcerting to see how much of the house was coming down. We were preserving practically nothing except the original form of the building. It was almost as if we had been given the house as a model and were reconstructing it from scratch, with a few internal changes in plan. The arguments against preservation of the original materials were mostly overwhelming economic ones. Time after time, we found that it would be faster and less expensive to fix a deteriorated finish or failing structure by taking it down and starting fresh with new materials. We could not be conservationists and expect to finish the rehabilitation job in a matter of months, or within any reasonable budget; that kind of project requires huge subsidies.

128   Wherever we could, however, we did try to

reproduce the quality of materials and workmanship that had been used in the original building.

New Framing

By the middle of August we were progressing in our work on the roof, the major structural jobs, and the interior of the Main House Unit. Our achievements included a large amount of new structural work in addition to structural repair. Footings for the new firewalls between the Woodshed, Icehouse and Barn Units had been poured, including foundations for the new fireplaces and chimneys they would incorporate. The concrete block firewalls had gone up, and the basic boxes of the fireplaces had been formed; ▼ later, our mason would line them with firebricks, install the flue dampers, and face the chimneys with stone. New framing was beginning to appear for the rooms of these units, and some of the new window and door openings had been roughed in.

The biggest changes could be seen in the Barn Unit, where we had conquered the foundation-wall problem and moved on to the floors. The dirt basement, or crawl space, had been excavated to a deeper level, and a new concrete surface poured in over plastic sheeting that served as a moisture

barrier. ▼ Individual new concrete footings were put in at the same time to support the different floor levels Jock had planned, and to anchor four new columns. These columns were going to hold up the ends of two very deep, handsome fir beams, new mid-span supports for the existing ceiling joists under the hayloft. The joists had originally spanned the entire thirty foot width of the barn without reinforcement, and although they had not given way over the years under the weight of countless hay bales, modern codes do not permit so long a span. Jock turned the Building Department's requirements into a pleasant architectural feature of the new

Barn by designing the exposed fir columns and laminated beams so they would define the living areas below, while they also propped up the old ceiling. At the same time, however, the rooms seem continuous, since the frames are not closed in.

This is one of the glories of Jock's plan for the Barn: you can see from one end of it to the other, through the greenhouse and the entrance hall, and vertically, up through the new cuts in the hayloft floor and entrance hall ceiling. The unit will still feel as open and spacious as the original barn even though it has a very private second floor. ▼ As we were working, it felt even more spacious

than the original barn, since we were just getting around to putting in the new floors and walls.

In order to support the ends of the floor joists below the level of the original high foundation wall (where it still existed), we framed in crib walls next to the stone. This simply provided a ledge for the joists to sit on. Once they were in place, we could lay the plywood subflooring and begin to frame in the walls. We had already cut out the section of the loft floor that would create such dramatic space ◀ and light at the west end of the living room, and the moment came when the first fir frame was erected. ▶ This was practically a piece of cabinet work, since the joint of the column and beam had to be carefully notched to fit perfectly, and its arrival really made us feel as if we were getting somewhere. We were moving beyond rough carpentry to all the attendant mechanical installations and to new features like skylights, balconies, staircases and partitions. All the units were getting so close to being ready for finish work that we were already shopping for hardware and appliances.

At this stage, the administration and co-ordination of subcontractors becomes a full-time job, as many owners sometimes learn the hard way. You will need a telephone on

the job site and someone to stay at it tracking people down and taking messages. If you do it yourself, plan to spend hours every day. It's not a job for the faint-hearted; the job's progress will depend completely on how well you can keep on top of the different representatives and tradesmen. Because we were doing this project for television, wheels tended to turn more easily than normal. As a private individual, you will find it a challenge to get each contractor on the spot at the right time and in the right sequence. You are juggling them, and each one of them is juggling you and all his other clients. It can be a very frustrating experience.   133

Chapter 7

Mechanical Systems

Before you can sit down to select bathroom fittings and lighting fixtures, your old house must be equipped to receive them. Although mechanical systems are usually hidden behind the walls and forgotten until they malfunction, they really deserve more attention. To start with, they usually account for thirty percent to forty percent of the budget in any job, whether new construction or rehab. Beyond that, they account for almost all the costs of using the house once it is finished.

People not familiar with construction tend to think of a house primarily in terms of its shell and surfaces—foundations, frames, walls, roof, floors, and so on. It is true that these elements account for most of the weight of a house and serve to define its basic configuration, but, by themselves, they are only the skeleton of the modern dwelling. It is much easier to understand the components of construction costs if you think of a building not as a single physical structure but rather as a series of systems. Think, for example, of the human access system—all the doors, halls, stairs, and so forth. We tend to take doors for granted, but they are often expensive and time-consuming to install, particularly in an old house. Other systems are often taken for granted: hot and cold water supply, waste-water disposal, electrical supply, air conditioning (including heating, cooling, filtration and general ventilation), lighting, and auto access. A kitchen is essentially a food preparation system. The list can go on: laundry, recreation, communication (including telephone, cable television, intercom and audio), security (fire and burglar alarm), storage, decoration, and so forth. While some of these systems are only conveniences, most have become an integral part of twentieth century life and housing. Think of your house, then, not as a container, but as a complex tool that enables you to accomplish many of the tasks of modern life.

Mechanical systems are generally quite straightforward. It is easy to get an idea of what equipment you'll need and why by going over the major systems in a residential building one by one.

Water

Plumbing is really two systems: a water system that supplies domestic hot and cold water, and a drainage, sewage and venting system that removes the used, contaminated water. *Plumbing* usually refers to the water supply system; the drainage system is subdivided into a sanitary system for the interior and a storm system for rainwater, both of which perform the same *DWV* (drainage, waste and vent) functions.

The parts of your supply/drainage system that should concern you first are the two ends—the water source and final waste disposal. If there is a major problem with either of these, your house could turn out to be uninhabitable without a huge injection of money. What happens in between the two ends can generally be fitted into your budget one way or another.

Water Supply

Every building requires a source of fresh water, which in the past was often supplied by individual wells and is now usually piped in from municipal reservoirs. If your water is supplied by town conduits, you are unlikely to face any serious problem, but there are still several things to consider. Although the Board of Health presumably keeps a watch on the purity of the water, you might want to investigate what kind of treatment occurs before the water is delivered. In addition to those chemicals added at a water-treatment plant, you may also be receiving harmful substances that have leached from the town pipes, especially if some of them are made of lead or certain older plastics, rather than cast iron. Since lead conduits were not generally used after the first half of the nineteenth century, it is unlikely that you will encounter a lot of lead; however, it was often installed until recently as the supply line linking a house to the town main. Theoretically, lead pipes oxidize to form a relatively impervious and harmless crust on the inside, but unfortunately water sometimes does pick up a poisonous amount of lead. Beyond this, it seems that municipal water supplies in many areas are being contaminated by various industrial pollutants. For all these reasons, it is not a bad idea to have a full chemical analysis of your water done before you purchase. There's not much you can do about contaminated water except not drink it. Hard water can be softened, tiny particles can be filtered out, and some minerals can be removed or neutralized, but in general you are stuck with whatever comes dissolved in your water.

Another important consideration is the volume of water supplied. Where drought and water shortage are a problem, the intake of town water may be limited by local regulations, so check to make sure the town can and will supply enough water for your planned re-use.

The main supply line from the town conduit to your building is your responsibility, because it's on your property. Be certain that it is buried deep enough underground that it won't freeze even during the coldest winter (depending on soil and climate, frost may penetrate the ground to a depth of between two and seven feet). It pays to be doubly sure about this, because houses are virtually uninhabitable without a water supply. The winter we were working on "This Old House" in Dorchester was exceptionally cold, and the main supply *did* freeze, leaving us without water and heat until spring came and the ground thawed out enough to replace the pipe.

If you buy an old building without a plot plan showing the location of utilities, you may have a little trouble finding the main water supply or sewage at a distance from the house. We were faced with this problem in Newton, but it turned out that Norm is a dowser, in addition to his many other talents. He demonstrated his technique for us, holding two thin copper rods parallel to each other and to the ground while walking around the area where underground water might be. As he approached water, the copper wires swung away from each other until they nearly formed a 180 degree angle. Norm couldn't explain it, but it certainly worked over the Newton water main. It doesn't appear to take gifted hands or brain waves either; the trick worked for me as well as for Norm, and I'm a born skeptic.

Once you've got yourself oriented, town zoning and sanitation regulations will tell you how many and what kind of clean-out access and fresh-air intake locations you must have for the supply and sewage pipes. One of the things you will need to have is an emergency shut-off valve on the main supply line, *outside* the building. At the Bigelow House, which is on Newton's water line, we put in separate hatches in the courtyard to house each unit's meter and emergency ▼ shut-off valve. These new hatches also give owners individual access to their sewage lines, which is better suited for repairs than the original city sewer manhole off to the east of the old courtyard entrance.

137

If your building is to be supplied by a well, the first thing to check is water purity. Have the Board of Health test a sample of the well water (or ground water, if you are putting in a new well). Once contaminated, a well cannot be purified, and generally remains polluted indefinitely, even if you remove the source of the problem. Sometimes contaminated water can be treated to be usable, and sometimes it is possible to drill or drive a new well on your land which taps a deeper and cleaner source of water. As a rule, though, any house where signs of contamination turn up in the well water will be a risky investment.

Assuming the water is clean, you should make sure that there is enough of it. Many old wells are not very deep, so again, you may be able to obtain a more dependable supply by drilling deeper, getting cooler and sometimes better-tasting water at the same time. Well-water must be pumped to the surface and stored in a holding tank; the capacity of the well may determine how much storage capacity you need and how often the pump will operate. Check to see what kind of pump you have, how old it is, and its rate of supply. If you have doubts that your water supply will be adequate for your needs, it would probably be worthwhile to consult an engineer.

Sewage Disposal

Just as any problems with your water supply may be unpleasantly expensive to solve, correcting inadequacies in the sewage system can also absorb a large chunk of your budget. For example, because some municipal waste-water systems are already working at or above capacity, you may find a moratorium on new bathroom installations, obliging you to provide your own sewage-treatment facilities or to live with the existing number of bathrooms. If your building is not connected to town sewage lines, you should investigate the condition and capacity of the existing septic system.

Unless the existing system has been installed within the last thirty years, it is likely to consist of one or more cesspools, or porous pits where the raw sewage collects, partially decomposes, and seeps into the surrounding soil. These have two main drawbacks: they are not considered sufficiently sanitary by modern standards, and they tend to fill up with sludge fairly rapidly. In the period when they were used, it was common practice to dig a new cesspool as needed rather than try to clean out an old one, so you will often find a large network of them on an old property. Although not up to modern standards, they were reasonably effective when used in this way.

Modern septic systems provide a tank designed to permit anaerobic bacteria to digest solid waste material fully, producing a compact sludge that needs to be cleaned out only every two or three years. Liquid waste passes to a seepage pit where it is further purified by bacteria and returned to the soil. Drainage *fields* are horizontal dispersal systems that can be used where the groundwater level is too high for a seepage pit. The aim of a pit or field is to disperse the treated liquid waste slowly into the soil well above the water table so that it has a chance to be cleansed fully as it filters down.

There are several things to watch out for in considering your septic system. You will probably wish to replace cesspools with a modern tank in the near future, and the new installation must conform to present sanitary standards. The most important of these is the "percolation" requirement, which specifies that water must drain downward through your soil at a satisfactory rate throughout the year. This means that if your land is low and subject to flooding, or marshy, or if the soil has a lot of clay in it, you may run into serious trouble with the Board of Health, and all your re-use plans may be jeopardized.

If your property has a septic tank rather than cesspools, you are less likely to encounter percolation problems, since the Board of Health presumably approved its installation. In the past, however, septic tanks were made of galvanized steel, and their life-spans are limited to around twenty years. Therefore, when a real-estate agent tells you that a modern system was "recently" installed, find out exactly how recently, what kind of tank it is, and whether there is proper outside access for clean-out. In the early days, this last feature was often omitted, and any tank which has not been cleaned out periodically is unlikely to be doing much good. Even if your existing system is in perfect condition, increases in bathrooms and occupants may force you to install a new system or additional capacity in your existing system. One small consolation is that modern concrete septic tanks do appear to last indefinitely, so you can expect your investment there to continue paying off for years to come.

Once you have satisfied yourself that water can be supplied and sewage disposed of at a reasonable cost, you should consider the internal sections of the system.

139

Holding Tanks and Hot Water Heaters

If your water is supplied from a well, a fairly large volume of it will be stored under pressure in a holding tank as soon as it enters the house, both to minimize the amount of time your pump must run and to ensure a steady supply of water even under heavy demand. Old holding tanks may be made of stone or cast iron, but you will generally find galvanized (zinc-coated) steel or iron. These do deteriorate: over the years, water slowly dissolves the zinc coating, finally exposing and rusting through the steel or iron underneath. You can sometimes get an idea of what shape your tank is in by tapping it with your finger or a small tool. If it rings clearly, that's usually a good sign, but a dull sound suggests heavy deposits, probably of rust. If you have to replace your holding tank, your plumber should be able to give you the most up-to-date advice on the kind and capacity of tank best suited to your water conditions (fiberglass ones are now widely used). At the Bigelow House, we didn't have to worry about this question since water came directly from Newton's reservoirs, which are, of course, "holding tanks" for the whole town.

Unless you are planning to re-use a tall building, water stored under pressure in the basement or coming in from a municipal line will circulate upward through the house without further pumping. In cases of very tall buildings, however, it may be desirable to pump water up to a holding tank on the roof so that gravity can provide for its circulation. If your building does use a roof tank, be sure to have it inspected by a competent building engineer, because it can place enormous loads on the rest of the structure.

The pipe leading out of the holding tank or main water supply line will divide into two branches, one for cold water and one for hot. If there is no main cut-off valve just before this branch, and if you plan to keep the existing plumbing, it is an excellent idea to have one put in, so that you can turn off the supply of water below the tank. Otherwise, you may find yourself unable to prevent a flood when a pipe springs a leak. If your water contains fine sediment or rust particles, this is the place to install a filter. Modern filters can remove extremely fine particles at relatively little cost or inconvenience, not only making your water look better, but also reducing drips in faucets and toilet valves.

The hot water supply branch leads directly to the hot water heater, and from there to the various hot water faucets throughout the house. When you are down in the basement looking for the hot water heater, do not be surprised if it seems to take the form of your furnace. It was popular not long ago to install furnaces containing what is known as *continuous-feed* water heaters. In this design, when you turn on a hot faucet, the flow of water causes the flame in your furnace to be ignited. The flame heats a coil through which the water is passing, while a thermostat measures the resulting water temperature and adjusts the size of the flame to achieve the desired heat. The advantage of this sort of system is that you can't run out of hot water as long as the flame stays on and the water keeps running. Also, you heat only the water you need, and when you need it. In theory, continuous-feed heaters should be more convenient and economical than a conventional tank heater, but in practice they have a tendency to produce erratic water temperatures, and are actually as a rule much less energy efficient than tank heaters. In a traditional tank heater, a large volume of water is gradually heated to the desired temperature, held at that temperature until needed, and replenished when used. In sharp contrast to continuous feed, almost no energy need be wasted during the gradual heating process, and with proper insulation

of the tank, heat loss during storage turns out to be relatively low.

In a tank heater, the size of the tank will depend not only on the needs of the building, but also on the heater's *recycle time* (the time needed to bring a cold tank of water up to the desired temperature), which in turn depends on the source of heat. For example, an electric water heater may have to be twice as large as a gas-heated one, because an electric resistance coil takes much longer to heat the water than a gas burner does. In a solar system, where all the water you will need throughout the day must be heated during those hours when the sun is hottest, an even larger tank is necessary.

At the Bigelow House, we planned to install gas, electric, and solar water heaters. The Woodshed, Icehouse and Barn Units all had the same requirements for hot water—two and one-half bathrooms, a kitchen and a laundry. Incidentally, washing machines use far more hot water than dishwashers, tubs or showers. In the Woodshed, we were going to use a forty-gallon gas heater, in the Icehouse an eighty-gallon electric heater, and in the Barn, where we decided to try a solar/electric combination, we would need a full 120-gallon tank. Larger tank sizes mean a somewhat higher initial investment; in the

141

It is not unusual now in New England to see the addition of a few solar collectors to the roof of an old house provide energy for hot water heaters.

long run this will not be as important as the cost and availability of the energy source. In the Main House, which has three and one-half baths, we used a fifty-gallon gas heater,▼ and for the Stable Unit, which has no laundry and only one bath, we planned an eighty-two-gallon solar/electric combination.

The solar system we planned to use is a closed liquid system. We could put the collectors on the roof of the new garage, where' they would soak up as much heat from the sun as possible. By pumping an antifreeze solution through them, then down into the hot water tank and back, we could transfer

142

the heat to the water. The system is *closed* because the antifreeze solution is sealed in copper tubing and just keeps recycling up to get hot in the collectors, then down to get cooled in the water, then back up again, and so forth. In an open system, the water itself would pass through the collectors, and although this might be somewhat more efficient, in the winter the water might freeze overnight and ruin the collectors. Whenever the sun cannot keep the water in the tank at 130 degrees F. (or whatever you decree), an electric coil comes on as a booster. The tank itself is lined with hydraulic cement, which stores additional heat, and is heavily insulated to keep heat loss to a minimum. On the subject of insulation, although most new water heaters are well insulated, older ones may not be, and you can often save a substantial amount of money by wrapping fiberglass insulation around the tank of your old heater.

Domestic Water Supply to Fixtures

From the hot water heater, a domestic hot water line parallels the cold water line to all faucets and showers, and to any other appliances that use water. The only divergence is that hot water is not supplied to toilet tanks or outdoor hose bibbs, while cold water is generally not supplied to a dishwasher. In some deluxe systems, the hot water line circulates in a continuous loop between faucets and the hot water tank, guaranteeing that hot water is instantly available at every faucet. Most older systems do not have this feature, which is just as well, since it can add considerably to operating and maintenance costs.

Finally we get to what most people think of as the essence of plumbing: the pipes.▼ If you don't plan to change the location of existing bathrooms, the old supply lines may serve you without repair for another fifty years, but it's equally possible that you might have to replace them at once. The first thing

to find out is what material they are. Galvanized or black iron were popular choices in years gone by because they were inexpensive and readily available. The drawback is that they eventually rust, resulting in discolored water, dripping faucets, and leaks. The most serious problems with galvanized pipe occur where it joins brass or copper fittings. An electrolytic reaction takes place between the different metals, rusting the iron much faster than usual, clogging up the pipe, and often so weakening the joint that it actually gives way.

Because of iron pipe's predictable mortality, you should consider replacing it while you are doing other work, especially if you will be opening up the walls anyway. Copper pipe, on the other hand, can last practically indefinitely under the proper water conditions, so if you find a copper system that seems in good health, there is every reason to believe that it will continue to serve you through the next century.

If your new layout, or the bad condition of existing supply pipes, forces you to install new ones, there are several materials to choose from. Galvanized pipe, though still available, is probably not a good idea, both because of its limited life span and because it's difficult to install. Copper, probably the most tested, durable material, is not hard to install, but is also the most expensive.

Recently, plastic pipe has begun to be used for domestic supply as well as waste lines, and may soon become the preferred material. In general, plastic is cheaper, more resistant to water acidity, and easier to install than copper. However, there are a large number of different plastics available for this purpose, and none has been in service long enough to have proved itself thoroughly. Among the possible choices for cold water supply are *PVC* (Polyvinyl Chloride), *ABS* (Acrilylonitrile-Butadiene Styrene) and *PE* (Polyethylene). In addition to these, there are two plastics safe also for hot water supply: *PVDC* (Polyvinyl Dichloride) and *PB* (Polybutylene).

One very major constraint on the use of plastic pipe is that many local building codes do not yet permit its use. Check with your plumber or with the local Building Department about this before making any decisions. As for choosing among the plastics available, you can either rely on the advice of your plumber (which will probably be to use copper), or you can talk to manufacturers, suppliers, other users, and even industry experts to get an idea of what the latest information is. Don't be afraid to use plastic supply pipes, but do keep in mind that nobody yet knows which plastic is the best in the long run and why. At the Bigelow House, we followed our plumber's advice and used copper for all the water-supply pipes.

The way pipes are joined together varies with the material. Iron pipes are threaded at the ends, treated with a sealant, and screwed into a coupling. Copper joints, on the other hand, are soldered—quite a beautiful job. ▼ The end of the pipe is first polished and covered with flux, then slipped into a sleeve coupling. The joint is heated with a torch while solder is melted around the edge and drawn into the joint by capillary action. Voilà—a perfect joint, almost like silvercrafting! You do have to be very careful with your torch, however. Copper conducts heat extremely well, which is why it is used in cookware, and you can start a fire if the pipe is in contact with wood or other flammable material. Plastic pipes are usually glued together, although they can also be joined in other ways.

Valves, Controls, Insulation

No matter what kind of piping is used, there are several features you should include in your system for efficiency and ease of operation. An adequate number of shut-off valves is first on the list. Professionals recommend having a valve on every vertical riser, on

each branch serving a bathroom or kitchen, and on the individual line to each fixture. This may seem like an awful lot of valves, but when a pipe bursts or a faucet breaks, you'll be glad to have enough controls. At an old farmhouse where a friend of mine lives in upstate New York, a galvanized-to-copper joint under a sink gave way at five-thirty one Saturday afternoon. The old system didn't have a local shut-off valve, so he had to go down to the cellar and turn off the main supply, which left him without water until he could get a new joint at the hardware store. Unfortunately, it was Labor Day Weekend and he had friends staying with him. . . .

Another feature to consider is the placement of storage valves at all low points in the system, as well as before outdoor taps and on any water tanks. Pumps, meters, filters and other equipment that may need servicing ought to be easy to disconnect, so it is advisable to install them with *unions* (bolted butt joints) ▶ rather than *couplings* (inserted joints). Valves should be installed around such equipment so that it can be removed with minimum disruption of the system. Capped air chambers may be distributed throughout the supply lines, particularly on any long straight stretches of pipe, to prevent the noisy hammering which can occur

when a faucet is turned on or off rapidly.

Insulation is another improvement to any plumbing system. Half an inch of fiberglass covering with a vapor barrier surface will not only reduce heat loss as hot water travels from tank to tap, but will also perform the important function of preventing cold water pipes from *sweating* over the summer. Sweating occurs during hot, humid weather, when water vapor in the air condenses on the cold surface of the pipes, and then drips onto surfaces below in surprising quantities, often with harmful effects. Also, hot and cold water lines should be separated by at least six inches to prevent heat transfer,

but in many older buildings they are considerably closer together. Insulating the pipes will help keep your cold water cold and hot water hot.

Sanitary Drainage

Supply pipes bring water to the faucets and other plumbing fixtures; then, after it has been used, the second half of the system, commonly referred to as *DWV* (Drainage, Waste, and Vent), removes it. The overriding function of the DWV system is to prevent contamination either of the water supply or of any part of the house by the waste water. This is the purpose of the *trap* that is located close to the drain mouth of every plumbing fixture. By creating a continuous water seal, a trap prevents any septic gases from backing up through the pipes and entering the room. *Vent stacks* (there should be at least one for every bathroom) protect the traps from the pressure which would otherwise siphon water out of their seals. Together, traps and vents make possible the high level of sanitation which we expect of modern plumbing, and which our health codes demand.

Aside from the job performed by the traps and vents, DWV differs from water supply in that water is simply drained off by the force of gravity rather than under pressure.

This imposes constraints on the way waste pipes can be installed. They are usually large in diameter, and, except in the traps, they must have a continuous downward slope. Vertical drops, of course, present no problems, but where waste pipes must travel horizontally,▼ it is important that they descend approximately one-quarter inch for every foot they traverse. If they descend less steeply than this, the waste water does not run off quickly enough. If they descend more steeply, the pipes tend to accumulate silt as the water courses down them. If waste pipes turn corners too sharply in the horizontal plane, they impede the flow of waste

147

water—where possible a series of forty-five-degree joints should be used instead of ninety-degree ones. In any case, the system ought to have a series of clean-out points at the bottom of vertical waste stacks, or in other convenient locations. As the waste water is about to leave the building, it should pass through a final trap to prevent extremely corrosive gases of the sewer from rising back up into the house waste pipes. If your DWV system lacks such a house trap, as many older systems do, it is wise to install one.

Choosing a pipe material for a DWV system is easier than choosing one for water supply. Plastic pipe has become so widely accepted as the best choice for this purpose that federal funding can be withheld from buildings that do not use it. However, some local building codes still prohibit the use of plastic, thus forcing you to pay much more for a comparable, or even inferior installation. You can choose between *ABS* (Acrilylonitrile-Butadiene Styrene) and *PVC* (Polyvinyl Chloride) for DWV applications. ABS is structurally stronger and less brittle, but may be a little less readily available. If you are prevented by local code from using plastic, traditional cast iron is the next choice. Cast iron is very strong and lasts forever, but it also takes forever to install, and is, as

you can imagine, extremely heavy. Copper waste pipes are unbelievably expensive, and unfortunately copper is very vulnerable to sewage gases. If your house lacks a house trap and has any copper waste pipes, check the tops of the pipes to make sure that no pitting or holes are present.

In order to demonstrate the various possibilities, we planned to use plastic waste pipe in four of the units and cast iron in the other. Copper is too expensive to justify putting in, even for demonstration purposes. Rich Trethewey, our plumber, was called on to do the honors. He let me help him with the plastic pipe, which goes to show how easy it is. The only trick to making a plastic joint is getting the two pieces in position before the glue dries.▼ The cast iron joinery was much harder, particularly the joints in the underground section. One end of each pipe has a *hub* or sleeve into which a normal diameter of pipe fits loosely. After two lengths of pipe are positioned correctly with the no-hub end of one inside the hub of the other, oakum (hemp calk) is packed into the joint in such a way that a ring-shaped cavity with an opening remains at the top of the joint. ▶ Molten lead is then poured into the opening, ▶ filling the cavity and making a permanent seal. A similar, less dramatic method is to use a rubber gasket sleeve,

which is secured and clamped with a metal cover, but this can deteriorate. The lead never needs replacing. Making copper joints is the same with the large waste pipes as it is with the little supply lines, except that you need a more powerful torch. In all cases, positioning waste pipes at the right slope and locating them in the exact place to meet fixtures is a tricky job.

Storm Drainage
The last concern of a plumbing system is the drainage of rainwater—usually a simple procedure, but nonetheless very important. We saw all too clearly at the Bigelow House

how serious the consequences of negligence can be.

For sloping roofs, gutters are the easiest way to collect and direct rain water. Designers who dislike the look of conventional gutters and down-spouts are constantly inventing ingenious modifications to avoid them or change their appearance, but the basic fact remains that water must be drained away from the house. One alternative is a gravel-filled trench located under the eaves-line of a sloping roof which is supposed to disperse the falling water into the ground, but this can be risky if there is a great deal of rainfall and the drip-line is close to the foundation walls. If the land adjacent to the house is paved, stone or brick channels can be worked into the pavement under the drip-line and made to work like gutters at ground level, carrying rainfall at a slope to a corner drain or seepage pit. Down-spouts from conventional gutters bring the water down to the foundation walls, where a splash pan may be enough to disperse it outward if the soil is porous and the rainfall not too heavy. Otherwise, it may be necessary to provide a drain pipe to carry the run-off away to a gravel pit for slow absorption.

Flat roofs can be drained centrally into interior drain spouts which parallel sanitary drainage pipes inside. This is a neat solution, as there will be only a few drain holes to keep clear of leaves. Also, the rain water can be piped directly to a dry well and eventually returned to the ground, or to a town storm sewer line. Storm sewers are kept separate from sanitary sewers so that the sewage treatment facilities do not have to process clean rain water.

The Bigelow House had especially high, steep roofs with very little overhang, so we were obliged to use gutters. Without them, the sides of the house would have turned into waterfalls during a heavy rain. We were determined not to jeopardize the expensive and painstaking work we'd done repairing previous water damage, so we did not balk at the initial cost or the large maintenance job of keeping them clean. Also, we felt that the profile of the wooden gutters had been part of the original design, and greatly contributed to the appearance of the house.

Except for gutters and down-spouts, all the water supply and drainage systems that we have been discussing will be installed and serviced by your plumber. His job, however, will probably not stop there; in many cases, he will also be responsible for your heating and air-conditioning. With this in mind, it seems logical to turn from water systems to a discussion of your house's heating and cooling system.

150

Heating, Ventilation and Air Conditioning

At first glance, the choice of possible heating systems may seem hopelessly complex, but in fact it can be broken down into three fairly straightforward decisions: what capacity system does your house require, what energy source is most desirable to you, and what kind of transfer system will you use?

System Capacity

The amount of heat your system must produce will depend on the heat loss of your building in the dead of winter. The calculation of heat loss is a lengthy but not very complicated procedure for which you can find full instructions in heating manuals and other reference works [see References]. The basic unit of measurement is *British Thermal Units* (BTUs) per hour. A *BTU*, like a calorie, is a measure of energy. It signifies the amount of heat required to raise the temperature of one pound of water one degree Fahrenheit, which translates into roughly the amount of heat produced by burning a wooden kitchen match. A typical single-family home in the average U.S. climate uses around 150 million BTUs per year. The size of your furnace, however, will not depend on the total number of BTUs you need, but on the BTUs per hour which your house loses during the coldest part of the year. Heat loss is a function of the area and conductivity of the building's heat-containing surfaces: walls, windows, outside doors, attic ceilings, roofs, ground floors, basement walls, etc., and also of their tightness. In a drafty house, a significant part of overall heat loss is caused by cold air seeping in from the outside. After estimating the losses in all these areas, and taking into account how cold the winters get, you or your plumber can arrive at a single figure for heat loss, and then choose a system which can supply that many BTUs per hour, or preferably a little more.

151

You should keep in mind, of course, that you can reduce the heat loss of an old building very dramatically by installing storm windows or double glazing, insulation, weather stripping, and sometimes caulking. As fuel costs rise, these things seem more and more worthwhile, especially where you can take a tax deduction.

In providing heat for the Bigelow House, one of our first jobs was to reduce heat loss to a minimum. We had already decided on double-glazing, but we wanted to make sure the rest of the house was as well-insulated as the expensive new windows. We were dealing with three different kinds of insulation problems: old construction, new construction, and attic or roof spaces. Because we were taking out all the old plaster, we didn't face the most common insulating constraint, namely the inaccessibility of wall cavities, but we still had to cope with certain irregularities, as well as the shallowness of the walls.

In the Main House, we found that a plaster coating had been put in over some of the inside surface of the exterior sheathing, sealing up one surface of the air cavity between the studs. Where we could, we left the plaster and filled the air spaces with blown-in cellulose insulation. This material is ground-up newspapers treated with boric acid, which acts as a fire retardant and also discourages insects and rodents from thinking they've found a pre-fab nest. We couldn't install the cellulose until the interior walls were plastered, because it must be packed into fully enclosed cavities. Two holes are made, one at the bottom of the cavity, and one at the top, and the cellulose is blown through a hose into the bottom hole at a fairly high pressure until it begins to over-flow out the top hole. Both holes are then sealed, and if the pressure and texture were correct, the cellulose will have been packed into all the crevices of the cavity so densely that it will never settle out.

This is clearly an ideal way to insulate without removing interior or exterior wall surfaces. We also used it because it gave us better insulation in the shallow spaces we had to work with than fiberglass would have; and second, because the cavities between the studs were often so irregular that it would have taken a lot of work to insulate them uniformly with fiberglass. Upstairs, and in a few places downstairs, we had room to add an inch of *EPS* (Expanded Polystyrene) foamboard above the studs and cellulose and below the plaster. Between the plaster and the insulation we put in a vapor barrier of plastic film to prevent humidity in the warm inside air from penetrating the insulation and condensing, damaging not only the insulation but also the wooden structure of the house. If you've pumped insulation in behind old walls, you can still achieve this effect by painting the interior surfaces with specially formulated vapor barrier paint.

In new construction, where the location of studs is regular and you can adjust the depth of the wall to suit your insulating needs, fiberglass batting is a very effective and efficient kind of insulation. Wherever we insulated newly-built walls, we used fiberglass bats. Not only are they easy to install, but they also come with vapor barrier built into their cover. You are taking care of two jobs at the same time when you staple them in.

Believe it or not, the most important single place to insulate in a house is the attic,▼ or roof. Because heat rises, the greatest heat loss in a normal house is likely to occur out the top. At the Bigelow House we were using all the attics as living spaces, so we insulated the roof cavities between the rafters, again using blown-in cellulose. If you will not be using your attic for living quarters, you may want to insulate its floor, so that you don't have to heat the unused attic along with the rest of the house. Even if you don't insulate any other area in your entire house,

the attic is worth doing. ▶

Caulking and weatherstripping are inexpensive expedients which will pay for themselves almost overnight. All outside doors should have weatherstripping to prevent cold air from slipping in, and any chinks in the armor of your house—cracks around window frames, gaps at the edges of junction boxes—can be filled easily with putty, caulking compound, or even with urethane foam spray available in aerosol cans. The technology with which you stop up holes is unimportant (old rags are traditional, though not as durable as modern materials), but each draft stopped in a chilly climate means a savings on your heating bill.

Once you have established the future heat-loss level of your house and determined the resulting furnace capacity, you are in a good position to calculate heat gain in the summer and the resulting air conditioning capacity needed, should you install a central plant. Again, manuals and technical reference books will give you full instructions for such calculations. Having determined what capacity of systems your house will require, you must next decide what kinds of energy source you will use to keep the air temperature where you want it.

154

Energy Sources

Almost all energy sources can be divided into things that are burned in the home or things that are burned elsewhere. The first category contains what you would normally think of as heating fuels—oil, gas, firewood, or even coal. In the second class are all the fuels burned to make electricity (radioactive fuels as well as oil, gas and coal). Also in the second class is the hydrogen consumed by the sun to give off its abundant heat and light. In the past, fuels such as peat and animal oil were used for heating, and in the future it may be possible to burn our plentiful supply of trash cleanly and economically. Only two sources of energy do not directly involve something being burned, namely hydro-electric energy (essentially generated by very sophisticated water-wheels) and geothermal energy, which is heat stored deep inside the earth.

Different energy sources will be more plentiful in different parts of the country. In selecting an energy source, you should carefully consider possible scarcities of fuel supply which you may experience over the life of your equipment, since it will be quite inconvenient if you ever find yourself unable to heat your house because of a fuel shortage. Unfortunately, questions of future supply of most materials remain anyone's guess.

The next question is one of cost. The first thing you must do is estimate how many BTUs per year it will take to heat your house. You will need to look up a *degree-day* figure for your region, and multiply it by a heat-loss factor for your house. One degree-day means that for one day, the outside temperature is an average of one degree Fahrenheit below sixty-five degrees (the temperature at which most heating systems go on). Again, a heating manual or technical reference will help you make this calculation. Next, estimate the average operating cost per BTU of each of the various sources of energy (obviously, solar will sound best at this point, being free). Manufacturers' specifications should tell you how much fuel or electricity has to be expended per BTU produced. You will have to rely on your prophetic powers to draw up a schedule of energy prices over the estimated life of the equipment. Now try to calculate the investment and maintenance cost per BTU for each installation you are considering. The easiest approach is simply to add up the purchase price, installation cost and probable maintenance expenses, then divide by the estimated number of years the equipment will last, and then divide again by the number of BTUs you will use each year. If you are feeling more sophisticated and daring,

155

you can try to include the opportunity costs too—that is, the interest you would be getting if you put the cost of the installation in the bank instead of into the house. In the end, you can add up the two types of cost and come up with a total estimated cost per BTU for each kind of installation. By comparing them, you can then see which energy source you have guessed will be cheapest. I favor this approach because it takes work and appears to be rational. Others maintain that since no one really has any idea what the future will bring, you should get the system you simply like best right now and depend on luck for the future.

Aside from availability and cost, you should also take into consideration technical characteristics of the various energy sources. For example, among the burn-at-home sources, gas heat is definitely the cleanest, while firewood does not burn cleanly and requires regular supervision. Among the burn-elsewhere sources, electric resistance heat is one of the least practical in most applications, but it has some advantages which may make it a possible choice for you. Electric resistance heat is supplied not by a central furnace but by individual units in every room, each of which usually has its own thermostat. A unit is an extremely simple and dependable piece of equipment, consisting, aside from the thermostat, of a heating element similar to one in an electric oven, surrounded by protective insulation and a layer of fins which dissipate the heat into the air. These units are quite inexpensive to buy and are very easy to install, because all they require is a heavy electrical cable—no piping, no ductwork, no lengthy thermostat communications. They are also completely clean in operation. The hitch, of course, is that under normal circumstances, the astronomical operating cost per BTU makes this kind of heat totally impractical in a cold climate. However, if you need only a very small number of BTUs per year, or if you are so cramped for room that you have no place to put a conventional furnace, then electric resistance heating may make sense.

A more efficient use of electricity is the *heat-pump*, a kind of large-scale backwards refrigerator or air-conditioner. Whereas a refrigerator removes heat from its interior and disperses it into the room, and an air conditioner removes heat from the room and disperses it out the window, a heat-pump brings heat from the outdoors into the house, even when the outdoors is much colder. This may sound a bit odd, but remember that the inside of the refrigerator is much colder than room temperature, and your compressor is still able to cool it. However, the colder the air out of which you are pumping heat, the harder and more expensive it is. When the temperature outdoors is around six degrees below zero Fahrenheit, most heat pumps just can't extract any more heat. When the outside temperature gets down that low, they have to depend on a resistance coil to provide backup heat until the temperature rises again and this becomes quite expensive. For this reason, they are just not economical in any climate where temperatures below minus-six degrees are commonplace.

The main drawback of the heat-pump has not been its operating costs, however, which in most areas of the U.S. are roughly comparable to burn-at-home fuels. Heat pumps have simply not been very reliable. As with any new technology, it takes a number of years to be perfected, and although it seems that recent generations of heat-pump are better than earlier ones, they have not yet been in service long enough to have proven themselves. Their largest advantage, on the other hand, is that the same piece of equipment can provide heating in the winter and air conditioning in the summer. The heat pump, after all, doesn't care in what direction it is pumping the heat, and can cool the house as easily as it can heat it.

In very old buildings, wood-burning fire-places were the sole source of heat. However, it should be remembered that what the fireplaces heated was generally the inhabitants, not the buildings themselves—as space-heating devices, fireplaces have never been very effective. Various improvements in their design, from Count Rumford's shallow plan perfected late in the eighteenth century to modern air-circulating devices, have increased their efficiency, but most of the hot air produced in a fireplace still goes up the chimney, often taking a good deal of warmed room air with it. Believe it or not, a deep conventional fireplace often cools a room more than it heats it! The best circulating designs combat this tendency by bringing in cold, outdoor air for combustion, and by passing room air over the back of fireplace surfaces heated by the flames, thus transferring the heat into the room before it all can vanish up the chimney. Although these strategies help a good deal, the fireplace remains more decorative than functional. Also, before you buy one of the steel circulating fireplaces, you may want to take into account its life-expectancy (very much lower than conventional firebrick).

Wood-burning stoves are a great improvement over fireplaces as space heaters. A modern, airtight stove, properly installed, can often radiate enough heat for an entire 157

Modern wood stove designs with closed combustion systems and room air circulation can deliver 70–75 per cent of the heat produced by the fire back to the room. An open fireplace returns only 5–10 per cent of its combustion heat and actually becomes a source of heat loss when the fire goes out.

By extending the hearth and inserting a metal heat deflector around the wood mantle, this decorative but inefficient fireplace has been made safe for a wood stove installation.

house. Many homeowners have achieved drastic reductions in their oil bills by supplementing their furnaces with such stoves. It is also possible to buy a wood-burning furnace, which disperses the heat more efficiently, evenly, and safely through your house than a stove would. Because of the extraordinary amount of heat they can radiate, wood-stoves can be quite a hazard and should be treated with great caution. The major drawbacks of wood-burning heat are that it needs constant attention, that the fuel is bulky, and that the combustion is not particularly clean. The major advantages are that wood is still a relatively cheap, abundant, and renewable fuel, and that many wood-burning installations have considerable charm.

There are several points to remember about burning wood. First, you want a stove that will heat the room, not send most of its heat up the flue. Second, you want a stove that will burn up all the tar and volatile gases in the firebox so that creosote does not condense in large quantities in the chimney and become a hazard. Combining these two requirements is something of a juggling act. For example, a stove with firebrick lining will maintain a high combustion heat in the firebox and be capable of burning up the tar and gases, but the firebrick will insulate the

158

stove walls so that they radiate less heat into the room. Although this can be compensated for by greater radiation from the flue, the cooler the flue air is, the more creosote will condense. Modern stove designs include baffles and secondary chambers to keep the hot gases in the stovebox for as long a time as possible, where they can burn fully and where heat can most easily be transferred to the room.

Stoves are rated by BTU capacity, like furnaces, and also by the area they are supposedly capable of heating. Most stoves on the market have quite a high capacity, so the main things you should be thinking about when you choose one are how thoroughly it burns creosote-forming tar and gases, what shape will best fit your needs, and of course, how attractive the stove is, since it will function as a piece of furniture in the room.

One thing to keep in mind about burn-at-home energy sources is that almost all require a chimney and a fair amount of space. Even more space, however, will be required by a so-called *active* solar system. A place outdoors must be found for the massive collectors where they have an unobstructed southern exposure and also for the storage medium, which holds the sun's heat until it is needed during the dark hours. This storage medium can consist of a small room

This two-family house in Berkeley, California, uses both active and passive solar systems for heat. Collector panels on the gabled roof provide heat for domestic hot water as well as the hot tub on the roof deck. In addition, the southern, vertical wall is made of concrete covered with fiberglass and acts as a heat sink for radiant heat.

completely filled with fist-sized rocks, or of several hundred gallons of water in a tank. Needless to say, without a large basement, storing such equipment can be difficult. A *passive* solar system, on the other hand, is built into the house, using a wall or floor both as collector and storage medium. While this can be achieved relatively easily in a new building, incorporating a passive system into existing construction is seldom easy.

Only where air conditioning is concerned is your choice regarding energy source fairly simple. Although in theory almost any energy source could be used, in practice most available systems appropriate for use in a residence are electric. If you live in a climate where you feel that air conditioning is necessary, this may also determine your choice of transfer system, as we shall note.

Heat Transfer Systems

There are three intermediaries used to carry heat from the place it is generated to the places it is needed: steam, water and air. Steam is used primarily in large buildings where extended distribution of heat is required, because at high temperatures and pressures it can carry large amounts of heat rapidly through pipes of small diameter. A normal dwelling, however, is not large enough to justify the added complexities and cost of a steam system, and unless you are renovating an old apartment building, you are unlikely to encounter one.

Most old houses rely on water systems to distribute their heat. The water involved is not connected to the domestic hot water supply, but is a separate system entirely, often containing antifreeze and rust preventatives. This fluid mixture is held in a reserve and passed into the boiler as needed. While there are many variations on boiler design, basically the water is heated either by circulating around the outer surface of a closed combustion chamber, or by passing through a coil, designed much like a radiator (intended to absorb rather than give off heat), located within the combustion chamber itself. A thermostat in the house turns on the burner when necessary, and the heated water is then pumped through pipes to individual *convectors* (radiators) and then back to the boiler. Looking at most old cast-iron radiators, it is easy to underestimate their efficiency. Many people assume that because they drip or sputter and look clunky, they should be ripped out of any rehab job, but in fact they are much more effective convectors than most of the fin-tube models you can buy today. With a little work they can be as good as new, and better than modern.

Not all hot water systems include a pump.   159

In Santa Fe, New Mexico, the addition of a passive solar collector, or southern-facing glass porch, to a traditional adobe house permits the owners to take advantage of free heat as well as a spectacular view.

In an old house, especially one that still has a coal furnace, you may encounter *gravity feed circulation*. Because heated water expands and rises, while colder water sinks, a system can be designed so that hot water leaving the boiler will rise up into the radiators, cool, and fall back through the return pipes, all without the aid of a pump. If you find unusually large heating pipes and a tank in your attic, you are dealing with gravity feed. They've been out of style for a long time, but they work very well with coal, and they may come back into general use, given the price of oil.

Piping for heat is different from that used for the water supply. If you are installing a new system, copper is most desirable in either application, but there the similarity ends. Plastic cannot, so far as I know, be used in a heating system, but galvanized iron is much less of a liability than it would be in a water-supply system. Copper and brass fittings that cause an electrolytic reaction with the iron can be avoided, and the water passing through the system can be treated so that it causes minimal rusting of the pipes. If you find galvanized heating pipes in place that seem in good condition, there is no need to remove them.

For an old water heating system to operate effectively, it is important to bleed air from the radiators on a regular basis. Because radiator valves are seldom completely tight, air tends to leak into the radiator during the summer or any period of extended disuse. Air in the system prevents water from circulating properly, so it is necessary to go around each fall from radiator to radiator bleeding air out of the little valve set in at each radiator's top corner for this purpose. Drips and sputters indicate worn or corroded valves, and are easy to fix by replacing gaskets or the whole valve if necessary. Loud clanking, known as *water hammer*, can usually be alleviated by introducing air chambers or other shock absorbers on branch lines. In general, if you can rescue the heat-transfer part of an old water system, it will serve you well.

Many solar systems use hot water as a transfer medium. Because water has a very high *specific heat* (i.e., it stores heat very well), it can double as a *heat sink*, or storage medium. Water passes through the collectors, is heated, and then held in a large insulated tank from which it can be pumped out to radiators as needed. Any water passing through the collectors should be mixed with antifreeze so that it does not turn to ice at night and cause a good deal of damage.

Water is usually not used as a transfer medium for heat pumps or cooling. If you imagine the effect of a cold radiator on a hot day, you'll see why. First of all, an enormous amount of humidity would condense on it, and you would need to provide a pan underneath. More important, the radiator would not move the cool air around, so you would end up with ninety-degree temperatures at the ceiling and fifty-degree temperatures at floor level. For this reason, the preferred transfer medium for cooling is air.

Air functions differently as a transfer medium than water or steam. At the energy source, it is treated much the same: it is passed over a hot (or cold) surface until it reaches the desired temperature, and then a fan distributes it through the house in ducts. But, unlike water or steam, air is not circulated in a closed system, but simply blown into the room. No radiators or convectors are needed, so hot air registers take up virtually no space, an obvious advantage. Concomitant disadvantages are that the air can be dirty, smelly, too dry or too humid. And the blowing process can be very noisy. Although very effective means of controlling all these problems are available, the equipment involved can be costly.

The ducts through which the air is moved are traditionally made of galvanized sheet metal. Fiberglass ducts are becoming increasingly popular because they are easier to install and quiet in operation. Air ducts are necessarily *much* larger than water pipes, having a cross-sectional area of two to three square feet or more. Finding a place for them in most old houses is usually difficult and sometimes impossible. The furnace should be centrally located and the registers should go under the windows at the periphery, so you are inevitably faced with the problem of hiding an extensive tree of horizontal and vertical ducts. They are too big to fit between most old studs or to be cut through framing beams, and you will be lucky if you can work out a scheme for hiding them in closets or stairwells. Usually, installing air ducts in old buildings means building chases out into rooms to enclose them. Because this procedure tends to spoil architectural details and dimensions, many people prefer hot water to air as a transfer medium in old buildings.

Heating the Bigelow House

To some architects, the inevitable question is asked so many times that it's become something of a joke: "How're you going to heat that space?" Well, sometimes it's a very tough question to answer, and we really had a job figuring out what to do in the Bigelow House. Not only did we have four units without basement or attic space for mechan-

161

The designer of this loft apartment in a recycled Chicago printing building made the most of large, exposed air ducts by incorporating them into the "roofline" of the kitchen enclosure.

ical equipment, but the units are of such varying size and layout that each heating system had to be designed differently. We wanted to use the most efficient and energy-conserving technology we could, and we found some exciting equipment on the market.

The breakdown of our mechanical solutions looks like this: in the Main House and Woodshed Units, which have the largest number of temperature zones to control, we used hot water systems because of their flexibility; and for the Barn, Stable, and Icehouse Units we decided to try new *heat-pump* technology. As an extra, we planned to install steel fireplaces with hot-air circulating fans back to back in the Woodshed and Icehouse.▶

Some of the modern equipment available for hot water heat is remarkable. In the Woodshed Unit, which doesn't have much of a basement, we installed an advance-design gas-fired boiler—right on the ground floor, in a closet! This was a real eye-opener for those of us who thought of boilers as noisy, bulky, vibrating, hot containers which can only be hidden away in a basement. This new model, called a "Heat-maker," is a technological by-product of government research into steam-powered automobile engines. It is so quiet, steady and

well-insulated that it can be installed anywhere, even on a wooden floor, with no offense or danger. Furthermore, it is so small that it can fit in a closet or under a staircase. It doesn't need a chimney for exhaust, only a five-inch vent pipe to an outside wall, so it can be used in apartments or in buildings that have no chimneys. It comes in a model with a built-in domestic hot water heater (the continuous feed type) or without, and we decided that in the interest of saving space that we would take a chance on the continuous feed operation. Everything else about the boiler seemed revolutionary, so we hoped the designers had not neglected

the bath water.

Even though it's small, this boiler is capable of heating an eight-room house. We were very impressed when the manufacturer explained why it works so efficiently. First, it functions as a *balanced system*, in which the intake of air for combustion is carefully controlled by an internal pump, and balanced by the rate of exhaust. It burns more economically than conventional boilers, where fresh air is pulled into a combustion chamber by the draw of exhaust gases up the chimney, which, in effect, fans the flame. Another reason for its efficiency is that the boiler heats only approximately one quart of water at a time. Therefore, when the flame goes off, the system loses only the amount of heat stored in one quart of water and the small, fourteen-pound copper coil. Conventional boilers usually heat six to ten gallons of water at a time, in much larger coils, so their standby loss, and your wasted heat, is considerably greater.

Air is drawn in from outdoors through a double-walled vent, which is really both an exhaust and an intake duct. An inner pipe carries out exhaust, and fresh air is taken in through the space around it. This air is mixed with gas fuel in a carburetor much like your car's, then burned in a small sealed combustion chamber. A finned copper coil

containing the convector water circles around the flame holder, so when the flame comes on, the water is heated almost instantaneously. Combustion fumes are fairly clean, since the mixture burns so effectively, and are exhausted (at 320 degrees) out the center portion of the vent duct. Because the exhausted air loses heat to the incoming air surrounding it, the exhaust doesn't leave the building at an incendiary temperature.

The water heated in the coil is directed out to whichever radiator in the house called for it. This is actually a second side to the flexibility and efficiency of our system. We found an assembly of Swiss-made radiators equipped with Swedish thermostat valves ▲

that allow the temperature of each radiator to be adjusted individually. Because each one has its own thermostat, it is connected directly to the boiler by a supply and return pipe. This meant spending a little more money initially for piping, but we felt that the advantages of instant response and individual adjustments made it worthwhile.

In addition to their superior function qualities, these radiators are quite beautiful—finished in colored enamel, they come in very flat versions that can be hung on the wall, or in baseboard units that can be installed in front of windows and still allow light to shine through. We put them in both the Main House and the Woodshed Units. We didn't put one of the heatmaker boilers in the Main House because we weren't sure it could handle all three stories, but we put in a similar gas-fired boiler with a copper fin heating coil and an outdoor thermostat that adjusts the heating temperature of the boiler to outside conditions. Since we did have a basement here, we had ample room for the boiler and tank hot water heater.

After we had been convinced of the efficiency and economy of these water-transfer heating systems, our real-estate advisers persuaded us to install central air conditioning as well. ▼ Boston is not a climate that de-

mands it, but apparently many apartment shoppers want it. This meant that in addition to the hot water system we would be putting in air ducts to handle cooling in the Main House and Woodshed. Clearly, this was redundant; we had to find a better solution for the other three units.

A heat pump was the single device that combined heating and cooling in an above-ground installation with energy-efficient technology. We could put the interior half of the pump in an indoor central closet, ◄ and locate the outdoor housings somewhere along the north or west walls of the units.◄ Since we were building new walls in all three of these units, we had no problem fitting in air ducts; in every case we could branch the ducts in the ceiling space between the first and second floors, feeding air down to the first story and up to the second. ►We'd had good reports from local homeowners who had tried these new heat pumps, and we hope they will do as well for us.

One final thing we did in an effort to cut down our energy consumption was to install old-fashioned ceiling fans in the Barn Unit. These run on about the same wattage as an ordinary light bulb, but the simple act of moving air around can create a temperature drop (in the summer) of eight to ten degrees. If it isn't very hot, just the move-

ment of the air can be enough to make you comfortable, without operating the heat pump at all. Similarly, in winter, the fans recirculate the warm air that collects up at the ceiling, thus creating warm drafts instead of chilly ones. We thought the fans were visually appealing, and considering the real energy savings they could provide, they seemed a good idea.

Both fans and heat pumps are electrically operated. In order to accommodate them and the other new electric needs, we had to rewire the Bigelow House, which meant wiring many places for the first time, a common part of almost all rehab or conversion projects.

The Electrical System
In an adaptive use project, the electrical system almost always needs replacing. For one thing, the demand for current will be vastly different, whether you're using a non-residential building for housing or putting more families in what has always been a home. Any new construction you do and occupancy changes you make will have to meet modern electrical code requirements, and that will usually involve upgrading the equipment you find.

The National Electrical Code (NEC) has been incorporated into the National Occu-

pational Safety and Health Act—in other words, it is law. Local codes may add special requirements, but the NEC is every American owner and contractor's guide to electrical installation. It is not a minimal safety guide, like the National Plumbing Code, but a high standard of provisions for every application. The only thing it doesn't stipulate is whether or not a homeowner *must* actually use electricity for power or what the source of electrical current has to be. You are free to burn kerosene for light and cooking without consulting the National Electrical Code at all. You can generate your own electricity with windmills, waterwheels or oil-burning generators, but the *way* you use it is governed by legal requirements.

It is not very difficult to work out what the basic electrical requirements for your project will be. An electrical system is analogous to a plumbing system, the major difference being that it is transferring power instead of water. The actual installation is very straightforward, once you know what fixtures and appliances you will use and where they go.

Power Handling Equipment
Like water, power is most frequently supplied by municipal or privately owned utility companies. Town power lines carry current at very high voltages to your property

At a windy site, this modern wind-powered electric generator is capable of supplying up to 60 percent of the electricity used by a typical American home. Utility power (or another generator) must make up the difference or take over entirely when the wind is low.

boundary, where your service line will tap into it. Between the main utility supply line and the service line will be a transformer that changes current from one voltage to another, or from transmission pressure to home-use pressure. A transformer is usually owned and maintained by the power company and can be located on a utility pole, set on a concrete pad at ground level, ◄ or even placed in a suitable place inside your building. Keep in mind that although they aren't terribly big, transformers make noise and get hot. This can be a problem indoors. Some transformers are oil-filled and present a fire hazard in a building, so if you have the option of putting one inside, make sure it's one of the dry-type units, or filled with a non-flammable liquid.

If you generate your own electricity, there is no need for a transformer, as power will be produced at the same voltage and frequency as it is used.

From the transformer, a service line leads across your property and into the building, where it joins a control panel board or *switchbox*, which contains the main house switch and circuit breakers for all the house circuits. This service line can often be laid ◄ underground, depending on factors like the type of terrain, weather conditions, voltage capacities, maintenance and service continuity,   169 and, of course, your budget. It is much more expensive to install electrical lines underground than to run them overhead from pole to pole, and repair service involves delay and expense for excavation, but there is an enormous advantage in terms of attractiveness, longevity of equipment and service reliability in bad climactic conditions.

The service line transports power in two wires, each carrying 120 volts, which can be attached individually or jointly to the separate circuits, giving your system the capacity for 120 or 240 volt circuits. Electricity has always been supplied like this. However, the size of the service line will determine how many *amperes*, or what volume of electricity is being supplied. (Electricity flows at a constant rate, so more of it arrives through a large wire than a small one.) Many old installations provided around forty amperes, while modern residential demand is for 80–100 amperes per home. It will almost certainly be necessary to replace the existing service line with a larger capacity one, especially if you are increasing the number of dwelling units.

Planning New Circuits

In addition to the volume of electricity you need, you will have to decide how to distribute it, or how many circuits to install. A circuit is a loop of electrical wire conducting power to a given number of outlets and switches, with a limited amount of consumption per circuit. Heavy-duty appliances like electric stoves, hot-water heaters, clothes dryers, air conditioners, dishwashers and garbage disposals are usually given individual, 220 volt circuits. Ordinary outlets and switches can be grouped on 110 volt circuits. The organization of numbers and divisions can be balanced most easily by the electrician making your installation, but you must be able to give him a complete plan of the light fixtures, switches and outlets for your new use. The larger volt wires are bigger and more expensive than ordinary 110 volt circuitry; if you can, plan your layout so that the service line enters the building where the run from the switchbox to the larger appliances will be short.

Get a professional's advice about whether old wiring and panel boxes can be re-used. Sometimes you can simply install additional circuits and transfer new loads to them. If the panel board (switchbox) has a main circuit breaker or fused switch, it can possibly be retained as part of a new service installation, but if it has been overloaded for some time, it will probably be in bad condition. The insulation on hidden connections may be unsafe. Of course it is stupid to throw away good wiring, but it is dangerous to continue using something that has deteriorated.

Planning switch and outlet locations takes a great deal of thought and imagination because it must be done before the walls and ceilings go up. There are code requirements for the occurrence and kind of outlets you must use, as well as for special signal devices like smoke detectors. In addition, there are many voluntary possibilities: *three- and four-way switches*, switched outlets, dimmer controls, convenience locations. The term *three-way* switch is misleading, as it refers to a light controlled by a switch in *two* locations (the "three" indicates the number of terminals in the switch mechanism); a *four-way* switch is actually a light switched in three locations. Most modern living quarters work very well with several of these switches, so that hall lights can be turned on or off from either end, garage lights from inside or out, etc. If you are trying to plan where to use them, take an imaginary walk through your house plan, as if you were returning from an evening out or going upstairs to bed. Where will you be entering and leaving a room?

You will need a switch at each place, so that you won't have to climb stairs, stumble across a room or try to find the back door key in the dark.

A kitchen probably has the most complicated electrical plan, but in your zeal for perfection there, don't forget outdoor and garage lighting, attic and basement lights, garden illumination, outdoor sockets, etc. And a further word to the design-conscious: if you do not tell your electrician exactly where you want each switch and outlet, he will position them wherever and however he is used to doing it. This will conform to code, but perhaps not to your taste. So, if you want to align the tops of the plates, or to group a number of switches together, or to install outlets horizontally instead of vertically, you must take the responsibility for specifying each location. It's well worth the trouble. You won't end up with random sprinklings of outlets in a wall or switches located so close to a door jamb that there isn't room for that beautiful old molding you've been saving.

Wiring and Circuit Breakers

Once you've made an electrical plan and identified the circuits, they will be installed with individual *circuit breakers* at the main switchbox. In older systems, fuses were

171

used to cut off power in overburdened circuits, but fuses are almost never used any more except in high-energy systems. A circuit breaker ▼ incorporates a manual switch and an overcurrent sensor like that in a thermostat. As current flows through the sensor, it bends from the heat, and if it bends too far, it releases a trip that opens the circuit and stops the current. The great advantage of a circuit breaker is that it can be reset with the manual switch; it doesn't self-destruct when it operates, like a fuse. Because of this convenience, you may want to replace old fuse boxes with circuit breakers; if you have to expand the capacity of your panel anyway

to include new and more powerful circuits, you may as well start with new equipment.

Choosing the type of wiring is another consideration. Most old electrical house wire was flexible, metal-clad cable, commonly known by the trade name *BX*. It is an assembly of two wires, each wrapped in rubber insulation, bound together with tape and enclosed in a cover of interlocking spirals of steel tape. It is still in common use today in a more modern form, with improved insulation and wrapping materials. One major disadvantage of the earlier versions of BX is that they didn't include a grounding wire or strip within the steel cladding, but relied on the cladding itself to act as a *ground* (an alternate path for current in case of damage to the wiring system or connected appliances). In old wire, the cladding will often have oxidized slightly at the joints, which will provide resistance to any current that might pass through it in a fault situation. In some cases, this resistance may produce enough heat to start a fire, or may not conduct electricity at all, leaving the human being at the other end of the appliance as the handiest receptacle for the current. If your electrician warns you against re-using old BX cable, he probably thinks it is unsafe because of deteriorating insulation or improper grounding. Modern

BX cable contains a separate copper grounding strip and can be used with all confidence. Because of its protective armor and flexibility, it can be snaked behind existing walls and under floor joists, making it very useful for rewiring jobs.

In small residential installations, electricians generally use a plastic covered wire assembly called *Romex* cable. ▼ This species also contains two hot (conducting) wires and one ground wire, each wrapped in plastic insulation and jacketed together in one flame- and moisture-resistant cover. By code, every cable has to include a minimum of these three basic wires. If you have

planned any arrangement of three- or four-way switches, the circuit cable connecting them will contain more wires for the increased number of terminals. Romex is easier to handle than BX cable but more vulnerable to physical damage, so BX is used more often in industrial situations and in superficial mounts where the wiring will be exposed to knocks and bruises. In fact, Romex is often only legal for residential installations of a relatively small scale—no larger than three stories.

All these wrapped conducting wires are made of copper. Aluminum is a cheaper alternative which is, however, only recom-

mended for larger supply lines where its light weight is an advantage. Aluminum wire doesn't match well with copper. The two materials expand differently when heated, so that the splices and terminals where they are joined tend to loosen. Also, aluminum oxidizes very quickly if it's exposed to air, forming a highly resistant film on the wire that can present a fire hazard. Because of these inherent difficulties, some localities have banned the use of aluminum cable in branch circuits, and when it *is* installed, it should be handled by an expert professional. Since these drawbacks to its application were only discovered after it had been used quite extensively for house wiring, it is not unlikely that you will find aluminum cable in an existing installation. If it seems to be in good shape, and codes allow it, you can re-use it, as long as you replace the switches, outlet receptacles and panel board (and, of course, segments of copper wire) with compatible material.

There are several brands of Romex cable on the market, all about the same price and quality. Because the Bigelow House was a residential installation with no more than three wired stories, our electrician used Romex cable throughout. The entire wiring system is broken down separately for each unit. The electrician planned on 100–200 amperes per unit, depending on the electrical load; those units with electric heat pumps obviously need more electrical capacity than the others. Each unit has a separate switchbox and meter, making it easier to regulate use and charges, and easier to isolate repairs. There is nothing particularly unusual about that part of the electrical system installed behind our walls. The individuality of the installation is reflected not in the components of the wiring system but in its planning—the decisions about where lighting and power will go.

We thought it would be a big asset to the general appearance of the property to eliminate telephone and electrical poles, so we buried the electrical wires. The lead wire from the main utility pole is aluminum, as is typical, but from the transformer on, everything is copper and completely new. Because we had taken down all the interior plaster finishes and were working with walls that had never been closed in, it was easy to install new wiring throughout. In fact, there were few places in the whole building where there had ever been adequate wiring for our purposes. Our electrician proceeded as if it were new construction.

In rewiring old construction, when you want to save existing finishes, you must use your ingenuity to make room for the new switch and outlet receptacles, new cable channels, and connections between floors. Openings are easier to make in plaster than in masonry walls, where every new space must be chiseled or drilled out; when faced with unwired brick or stone walls (for instance, if you are exposing a masonry bearing wall), you should usually expect to make a surface installation. No matter what the original building material, hiding a new cable is a challenge. There are various carpentry tricks, such as making a groove for electrical cables behind baseboards or picture moldings, and if you have access to an unfinished basement or attic, it may be easy to string cable around from those spaces. Otherwise, you may find yourself lifting floorboards and boring blindly through wood blocking in the wall structure.

Secondary Electrical Systems

While you are planning the electrical installation, don't overlook low-voltage wiring for smaller electrical systems like telephone cables, intercoms and doorbells, smoke detectors, thermostats, ▶ and burglar alarms. It is far easier to install these little signal systems while the plaster is down than to add them to finished walls. In addition to these amenities, there are various other luxuries such as cable television, central vacuuming,

hi-fi speakers, etc. These are more likely to be included in the planning for new construction: an old building is going to be strained enough with the inclusion of new plumbing and air conditioning. If you insist, you can impose extravagant modern technology on old buildings, but there will be an aesthetic price to pay. Whether or not a juxtaposition of high-tech and antique appeals to you is obviously a question of personal taste.

In talking to architects, contractors and other rehabbers, you will hear many different recommendations about technological renovation. Many will recommend taking out

every existing pipe and wire and starting over; others will say you must carefully rehabilitate and adapt existing equipment. Anyone who is concerned about preserving the character of an old building will agree that new mechanical installations should be sympathetic to the kind of renovation you're making. Obviously, if you're undertaking historical restoration, you will not want to install air-conditioning ducts and fluorescent lights in a new, lowered ceiling, but these same features may not be out of place in another kind of re-use. There is no way to generalize except to say that the choice of systems will significantly effect your rehab aesthetically, functionally and financially.

If you find a building without any mechanical system, your course is clear. The major question in that case would be whether the existing structure can accommodate all the necessary parts of the modern systems required by code for your new plans.

If original equipment exists, can it be adapted to your new use or do you need to remove it and start over? Consider whether the old equipment might be adequate for all or part of your requirements, whether it can be made compatible with new equipment, the ease of maintaining and servicing it, its performance in terms of pollution and energy efficiency, and its projected years of service.

Now that you are familiar with the workings of modern mechanical systems, you should find it easier to evaluate their impact on an existing building. Don't expect contractors to make the decision for you; they are not accustomed to looking at a mechanical problem from a preservationist or historically sympathetic point of view. You should rely on them to know the applicable code requirements and the quality of available materials. Together you will be able to design a mechanical renovation that comes closest to satisfying you and the Building Department.

When the mechanical work is complete, you are in the home stretch. There is nothing left but the finish work—cabinetry, wall, floor and ceiling surfaces, light fixtures, bathroom fixtures, appliances, and decoration. It goes more slowly than you expect at this point, but little by little you can see the finished house emerging.

177

Chapter 8

Surface Finishes

Although the structural members of a house hold it up and the mechanical systems make it work, most of what you usually see of a house, inside and out, is its skin. The outer skin has one common defensive function (keeping out wind and water), and it is generally divided into two distinct surfaces, the roof and side walls. Sometimes, as in the case of the Bigelow House, the entire exterior can be covered with a single material. Interior surfaces, on the other hand, are always as diverse as the huge variety of purposes they serve. You may be tempted to consider surface finishes only superficially important, but this is a mistake. If you think

of them as icing on the cake, remember that the rest of the cake exists to support the icing.

Exterior Surfaces

Because a building's most important defense against the weather is its roof, the first step in rehabbing—before any interior construction begins—is to fix up the roof, if necessary. The kind of repairs your roof may need will be determined by its surface material, which in turn depends on the slope of the roof. Different techniques, all with various strengths and weaknesses, are appropriate to steep, shallow and flat roofs.

Overlapping shingles, whether made of wood, asphalt, slate or tile, can only be used on a roof that rises more than one foot for every three horizontal feet (at an angle of more than eighteen and one-half degrees). If a roof is less steep than this, water will not run off fast enough and will tend to seep into the joints between shingles. Even on a steep roof, seepage can occur near the eaves if accumulated snow or a clogged gutter slows drainage.

A roof with a shallow slope may be too flat to take shingles and too steep for tar or asphalt, which tend to ooze downhill. The solution is metal. Traditional roofing metals are tin, terne-covered iron (terne is a mix-

179

ture of lead and tin), and copper. Sometimes you will find ordinary galvanized sheet metal used on old roofs, although it tends to deteriorate very rapidly unless kept scrupulously painted. Pieces of metal are joined together either with flat soldered seams or with standing seams, which may be soldered or simply rolled. Loose seams are the most common cause of leaks in metal roofs, although damage to a protective coating, followed by rust, can also lead to holes. Dramatic failure can sometimes occur when strong winds tear a whole section of metal away. If you must do extensive repairs on a metal roof, the most economical solution probably will be to replace it with shingles, provided it is steep enough to take them. If the slope requires metal, copper is the best but most expensive choice; you can comfort yourself that it will give you many, many years of protection.

Because flat roofs often have to support puddles of standing water, they must be strong and completely waterproof. A standard flat roof consists of tar or asphalt poured over successive layers of roofing felt and finally covered with gravel. There are a number of sophisticated variations on this theme, but the principle remains the same. Once in place and protected by gravel, flat built-up roofs are fairly dependable; prob-

lems tend to develop not in the roof membrane itself but around flashings and edges.

The side walls of most old buildings are clad in wood or masonry. Most masonry walls are naturally impervious to water. Brick and some sedimentary rocks, however, can lose their seal through erosion, and water can then seep in. Frost and sun can cause flaking or even severe disintegration. A common contributor to this kind of problem is excessive zeal in cleaning, so if you want to remove the grime of ages from an old brick or limestone wall, soap and water is far preferable to sandblasting or harsh chemical cleansers. All stone is subject to some wear from weather, pollution and climbing vines, but the weakest points of most masonry walls are the mortar joints. If you keep those in good condition, you have centuries to worry about surface deterioration.

Of the numerous types of wood sheathing you might encounter, the most common are shingles, board and batten, clapboard, and vertical butted boards. All of these should last indefinitely if properly maintained and painted, and superficial damage can usually be repaired easily without replacing boards. Caulking compounds are among the very few modern materials that are unquestionably superior to their historical counter-

180

When the Detroit Cornice and Slate Company sold its building for re-use, the new owners were eager to restore the emblematic eagle which was missing from the roof. Most of the other cast iron details were intact and could be restored with cleaning and fresh paint.

Even in a state of disrepair, the graceful cast iron and glass facade of the McLaughlin Building in Boston attracted the attention of rehabbers as a potentially wonderful place to live. The upper stories of the building are being made into apartments, and the ground story will continue to be rented for commercial use.

parts. The oil–based caulks of the past used to dry out, harden, shrink and finally crumble. In contrast, the silicon–based caulks used today retain their elasticity and sealing qualities for years.

Of course, rotten siding will have to be replaced, but this kind of work is not as expensive as you might think. Above all, don't rush to the phone to order aluminum siding just because your wooden clapboards look a little beaten up. Aluminum or vinyl siding is a great temptation to homeowners who want to reduce maintenance costs, but quite aside from their unmistakably twentieth century appearance, these non-porous ma-

terials can cause more trouble than they save. They can be applied in such a way that they trap water in the wall cavity, eventually causing the framing to rot. Wood siding *breathes*, allowing any moisture that leaks past the interior vapor barriers to escape. Of course, all wood (except cedar) must be protected on the outside by paint or a water-repellant stain. Cedar holds up so well against the weather that it doesn't have to be coated.

We became very well acquainted with all the nuances of cedar shingles during our renovation of the Bigelow House. ▼ Originally, the entire structure had been covered

by a continuous cedar skin. Shingle Style architects loved surface continuity, and Richardson probably made the Bigelow House roofs as steep as they are to emphasize their similarity to the walls below. When later owners reroofed the building with asphalt shingles, however, the effect was reversed—one of the first things we noticed about the house was how inappropriate and obtrusive the asphalt looked. On a house with a more retiring roof, it would not have clashed so obviously with the siding. We felt as strongly as the Historical Commission that the original effect should be restored.

On the side walls, the old cedar shingles were still more or less in place, but so many of them were decayed, split, waterlogged, or missing entirely that we decided to pull them off and start fresh. Our decision seemed all the more sensible when we found out how many of the walls around the Barn, Icehouse and Woodshed Units had to be replaced entirely.

Reshingling the roof was probably the second largest job in our rehab project, after the foundation and sill repairs. Our roofer, Jimmy Marcotte, and his crew started in during the first week of work to remove the old asphalt shingles. There were two layers of them: when the first had worn out, another had been laid on top. Now the second

layer was coming to the end of its usefulness. I was curious to see if we would find the original roof underneath, but what we uncovered was the bare sheathing over the rafters.

There are several reasons why asphalt shingles had been chosen as a substitute for cedar by previous owners. First of all, they are easier to get and cheaper than roofing-quality cedar, and second, they are much more fire-resistant. Fire is less of a hazard now than in earlier times, in no small part because modern roofing materials are so effectively fire-resistant. With all the chimneys, fireplaces, and woodstoves in our new plans, we were a little worried about going back to a wooden roof—we certainly didn't want out handiwork to go up in smoke! Our solution was to order top-grade quarter-cut cedar roofing shingles treated with a fire-retardant. Although the fire-proofing is expensive, we felt the added safety justified the cost. These shingles are cut from the center of the tree, so that the grain runs roughly perpendicular to their surfaces. As a result, they are less permeable and lie flatter than shingles cut from the edges of the tree, where the grain runs in long, shallow, horizontal curves.

Jimmy, our shingle specialist, estimated that we were going to need about one hun-

dred squares of shingles for the roof alone. A *square* is one hundred square feet, so our roof was going to require about 10,000 square feet of shingles. Add to this the ninety squares which Jimmy estimated that the side walls would need and you have quite a job! Jimmy recommended that we use a different grade of red cedar for the siding. Since neither hot cinders nor much rain and snow will fall onto the vertical shingles, they don't need to be as fire- or water-resistant as those on the roof. We did, however, get them quarter-cut so that they would match the ones above and because the straight grain is so much easier to handle.

Jimmy and his crew worked around the roof in sections, trying to keep as much of the roof as possible water-tight in order to minimize rain damage to the interior. This turned out to be a good idea; at one point, for example, roofing was unexpectedly delayed for a month while we waited for shingles, and if we'd been less cautious, a large part of the house might have been exposed to the elements.

Although shingling is basically a simple process, there are a lot of tricky details. You must always start at the bottom and work your way to the top,▼ since each succeeding course of shingles covers a good part of the

course below. Before you lay any shingles at all, however, you must seal the lower edge of the roof in some way to prevent water backed up by ice dams from seeping into the house, obviously a concern limited to cold climates. Ice dams are caused when heat from the sun or the furnace melts snow lying on the roof. Water runs down into the gutter and refreezes, gradually clogging the gutter and backing up onto the roof, where a standing puddle can then form. Some roofers don't even bother with shingles on the first three or four feet of a roof but install a wide strip of metal flashing instead. Being very anxious to avoid any breaks in the continuous shingle skin of the Bigelow House, we preferred to use a modern *bituthene* membrane especially designed to form a waterproof seal under the shingles. After first painting the roof sheathing with a primer to prevent any moisture already in the wood from causing problems, we rolled out a ▶ thirty-inch wide strip of the ice-shield at the lower edge of the roof. Bituthene consists of rubberized asphalt sandwiched between a film of polyethylene on one side and adhesive on the other. The adhesive bonds it firmly to the sheathing while you nail the shingles down onto it. The rubberized asphalt closes around the nails and maintains a perfectly waterproof seal, allowing you to

apply shingles right down to the eaves without sacrificing any ice protection. The shingles, in turn, protect the bituthene from drying and deteriorating in the sun.

It is important to lay the first, or lowest, course of shingles so that its bottom edge is straight. If you're not lucky enough to have straight sheathing to follow (and you probably won't be), you can use a string, straightedge or line-of-sight to stay aligned. Shingles are attached with one or two galvanized shingle nails each, placed half to three-quarters of the way up the shingle. A quarter-inch gap should remain between the edges of any two roofing shingles, not only

because they swell when wet but also to provide ventilation channels back into the areas covered by shingles above. Because the first course has to withstand the heaviest attacks of wind and ice, it is reinforced by a second layer of shingles laid directly over the first in such a way that every gap between shingles on the bottom is covered by a shingle above, and each gap in the upper layer falls near the center of a shingle below. With the initial double layer in place, you can then proceed to the second course. Although shingles are around eighteen inches long, each course overlaps most of the one beneath, leaving an exposed portion, or ▼

weather, no more than five and a half inches wide. This means that any point of the roof is protected by at least three layers of shingle. If the peak of your roof is not quite parallel to the eaves, as sometimes happens in old houses, you may want to vary the weather from one end of the roof to the other so that you end up with a uniform number of rows all the way along.

In our case everything was parallel, so once we had settled on a traditional five-inch weather, all we had to do was measure up five inches from the bottom of the first course and use a straight edge or *chalk line* to mark where the bottom of the second course should go. A chalk line is a length of soft cotton string impregnated with colored chalk. In order to mark off a line between two points, you stretch the string tight from one to the other, lift the center a few inches off the surface (like pulling back a bowstring), and release it. When it snaps down it leaves a chalk imprint in a straight line between the points.

The most important principle in shingling is to keep the gaps between shingles in any three consecutive courses as far from each other as you can. When water flows into a gap, you want to be sure that it lands on the impervious surface of a shingle beneath and is carried on downwards before it can seep

185

Decorative slates were often used to shingle visible roofs on Victorian buildings. If you can repair a slate roof by replacing a small proportion of the individual slates, it will pay off visually and economically.

sideways to another gap. You must be sure that at any point in the roof, water has to take as long a path as possible to pass through the three layers of shingles into the interior. If you adhere to this principle your roof won't leak—at least, not through the shingles.

There are a fair number of areas that shingles alone cannot protect. Around chimneys, dormer windows, skylights and vent stacks, for example, or in valleys where two different planes of a roof come together, you must resort to metal flashing. ◀ The best material to use is lead-coated copper, which will not oxidize and stain the shingles, as plain copper will. Unless treated with a clear acrylic sealant, however, lead-coated copper will lose its coppery look and weather to gray. Aluminum, although not as versatile or durable, can also be used. Flashings are generally laid over a base of felt roofing paper and nailed down at the edges.▼ Shingles are then used to cover as much of the flashing as

possible without inserting a nail into it. Exceptionally wide shingles are useful for this, particularly in a roof valley, where one side of each shingle has to be cut at an angle.▼

Not only did we have quite a lot of lead-coated copper flashing to install at the Bigelow House, we also had several areas that required metal roofs. One was the hipped roof of the sunbath, which had too gentle a slope to be shingled. We used strips of lead-coated copper painted with clear acrylic so that it will remain coppery in color. The joints between the strips are standing seams, which depend on gravity to keep water out of them and don't need to be soldered. They

187

are formed by turning up the two edges to be joined so that they rise side by side several inches above the level of the roof. Every foot or so, a clip has been nailed to the sub-roof ▶ and bent up between them. The whole sandwich is then rolled over onto itself and flattened out, so that the final seam "stands" about an inch above the roof. This is high enough that water prefers to run off rather than try to scale the seam.

The belvedere, which has to bear the weight of people standing and walking on it, needed a stronger, flat type of metal roof. Here we put down a number of layers of heavy felt padding first, and then laid 16-inch-wide strips of lead-coated copper. Our roofer had measured the roof in advance so that he could precut these strips to size in his shop and solder on nail clips to hold them down. We nailed on a hood strip as a starter at one edge, then interlocked each subsequent strip with the one before it as we hammered in the nail clips. Finally, we sealed the flat seams where strips overlap with lead solder. It's a tricky job if you're not a professional, but the result is very satisfactory.

Complicated as the metal roofing was, some of the shingling was just as demanding. The turret roof, for example, while steep enough for shingles, curves so much horizontally and vertically (like a witch's

A gold dome which catches the first and last rays of the sun was a traditional roof flourish on many old academic or ecclesiastical buildings. Gold-leaf is not a cheap roofing material, but invariably a cheerful one.

hat), that straight shingles just don't conform to it. You can solve the horizontal part of the problem by using only narrow shingles, tapering their tops, but we found that to avoid breaking most of the shingles on the vertical curve, we had to soak each batch overnight in water to soften them. ▼ the problem was that the fireproofing had made them more brittle than untreated shingles would have been.

Another area that required a little shingling artistry was the lid of each eyebrow window. As each course was carried over the window and emerged at the same level on the other side, it had to squeeze together

with courses above and below, narrowing its weather so that all the courses could still fit. The effect is exactly as though the roof had opened an eye, stretching and compressing its shingle skin. ▶

Among the hardest original details to reproduce were the little wing gables that swept out at the corners of the house to protect the ends of the gutters. In this case, because the vertical curve was so tight that even softened shingles wouldn't work, we had to cut *kerfs*, or lateral notches in the underside of each roofing shingle with a bandsaw. The kerfs left only a thin membrane of wood to bend, allowing the shingle to conform to the sharp curve without breaking. These kerfed shingles were as vulnerable to wind and storm as the rest of the first course, so we were careful to use two layers. Siding the gables was even trickier than roofing them. On the front and underside, we could simply use very short sections, but on the sides we had to notch out several shingles to conform with the intricate and attractive profile that had been a feature of the original house. In spite of Jimmy's expertise, it took him a couple of tries before he could get this detail right.

Shingling the side walls was almost as big a job as the roof, and it involved even more fancy details. One advantage was that since

This 1888 residence in Cambridge, Massachusetts, designed by Hartwell and Richardson (no relation to H. H.), has many "Shingle Style" features: turrets, gables, lattices, asymmetrical windows, and a continuous, undulating shingle skin.

we didn't have to keep the surface water-tight while we were working on it, we could remove the old shingles all at once. Some of the sheathing underneath was in bad shape. We would have used three-quarter-inch plywood to replace it in the rotten spots, but the original sheathing was seven-eighths-inch ledgerboard. Although plywood is stronger and easier to install, the one-eighth-inch difference between three-quarter and seven-eighths would have produced a noticeable irregularity in the surface, so we simply used new ledgerboard.▼ When the sheathing had been repaired and we were ready to shingle a given area, we first stapled up a layer of kraft

paper as a wind barrier. The paper is meant only to stop drafts from coming in; water and water vapor pass through it easily, allowing any dampness inside the walls to escape.

As in all shingling, we had to start at the bottom and work up. Using a level and straight-edge, we first laid out a uniform bottom line for the first course. Then we had to provide the *skirt*, sweeping out an inch or so at the bottom of the wall to divert water from the foundation. In order to form this curve, Jimmy used wooden props or ◄ *squinches* under the double layer of first course shingles.

On the side walls we weren't able to decide on a single standard "weather" width as we had been able to do on the roof. If the facade is going to look right, the tops of the window frames must coincide with the bottom of a shingle course. By measuring the distance between the bottom of the first course and the top of a window, you can calculate exactly how wide a weather you will need (within rough limits—in our case between four and a half and five inches) so that a whole number of courses will fill the space perfectly. Then, using a pole hanging from the tops of the window frames as a measuring stick, you can mark off with a straight-edge or chalk line where the bottom of each

course should fall. If the windows are all evenly positioned around the first floor, you clearly have to make the calculation only once. At the Bigelow House every window was a little different, and we had to plan each wall individually in order to match the shingles with the windows as neatly as possible.

At the top of the first floor, we had to reproduce some characteristic shingle ornamentation. Here, a second skirt, made up of two shingle layers just like the first course, curved out over a piece of trim molding. ▼ The edge of the bottom layer of shingles protruded about one-half-inch beyond the molding and was cut in a saw-tooth, or,   193

more technically, a dragon-tooth pattern. Our millwork supplier was able to duplicate the trim molding without difficulty, and the dragon teeth were easy to cut out on the ▶ band saw, so this detail was far from the most difficult we had to duplicate. However, this one small decorative touch, which a modern architect would never think of, seemed to set the tone for the entire facade.

Consistent with the Shingle Style emphasis on surface continuity, there had been no cover boards at the corners of the house. Instead, the shingles ran right up to and around the corners without a break. To avoid creating one single long corner seam into which water could easily seep, the overlap of the ◀ shingles alternated from course to course: first one side would cover the other, then the second side would cover the first.

There were a lot of fussy little wall areas that required time and skill to cover neatly. Probably the most challenging was the ▶ rounded underside of the Main House turret. Like the turret roof, this area curved vertically as well as horizontally. It didn't have to shed water particularly well, since the turret above protected it, but even so every shingle had to be carefully cut to size. In this case it was the weather that we tapered, not the upper part of the shingle hidden by the courses above.

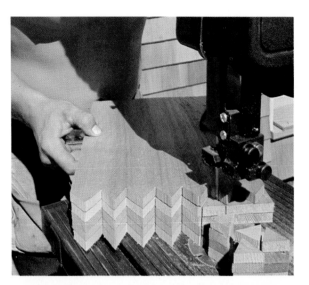

195

Our shingle job was certainly not typical of modern construction. Not only did we have a lot of unusual detailing, we also could have used only half as many shingles on the sidewalls if we'd made the weathers twice as wide, which still would have been adequate on a vertical surface. Our shingling crew needed over five months of hard, steady work to cover it all. We were glad that our commitment to the Historical Commission forced us to use the same narrow weather on the walls as on the roof, because the resulting continuity between roof and walls really was beautiful. We were less happy with having to stain the walls red to match the original house; ▶ we couldn't help thinking that the vast expanse of prime cedar would have looked superb as it weathered naturally. Our shingling crew needed over five months of hard, steady work to cover it all.

The last touch to the exterior skin was paint for the trim. We could tell from some sheltered areas under the bathroom of the Main House that the original trim had been painted a dark green color, and we assumed that the doors had, too. We found a good dark-green oil-based paint that would go ▼ well with the dark red stain on the shingles, and went to work. Even putting on two coats, it didn't take long; the outside skin was finished!

196

Interior Surfaces

Because of a delay in receiving the roofing shingles and because of our filming schedule, we found ourselves beginning the interior finish work at the Main House before we had completed the new roof. Even though this caused some concern about keeping rain and wind off new plaster finishes, it really was exciting to see the rooms take shape. Open areas populated by exposed studs, electric cables, plumbing and heating ducts disappeared so fast that we hardly associated them with the new spaces that appeared.

Before we could close in all the ceilings and walls we had to decide about insulation. In most of the Main House, we relied on loose cellulose blown into the wall cavities through holes in the exterior sheathing, after the interior walls were up ◄ but before the outer walls had been shingled. To insulate a ceiling, however, we used fiberglass batting, and in a couple of places where we couldn't fill the whole wall with cellulose, because of cavities for sliding doors or windows, we used one-inch styrofoam boards between the studs and the plaster.

We felt the most important ceiling to insulate was the one between the second and third floors. The third-floor bedrooms, situated directly over the master bedroom, would most likely be guest or children's rooms. As guest rooms, you wouldn't want to have to heat or cool them when they weren't being used, so you would want them to be insulated; as children's rooms, you would want them insulated acoustically. Somehow, people imagine that old houses are built so solidly that you can't hear a thing through the walls or floors. It's true that plaster cuts down noise transmission better than modern sheetrock, but a floor can transmit sound much better than you might expect. Fiberglass batting could serve as insulation for both purposes, in addition to being easy to push up between the joists

before the ceiling went in, so we had no hesitation about using it.

We used styrofoam board in the master bedroom and the living room below it because of the sliding door and window cavities, which prevented us from completely ▶ filling the walls with cellulose. Foam board is a good solution where the cavity between the inner and outer walls is so thin that adequate insulation cannot be obtained by filling it, or where the cavity has to remain open for some reason. It has the advantage of being rigid enough to be nailed over the studs to thicken a wall, and also able to provide a lot of insulation in a small space.

In every wall there should be a vapor barrier. In the winter, the warm air inside the house carries a lot of moisture, which would condense if it could get to the cold surface of the outer siding. Without a vapor barrier, you can end up with a constant small waterfall inside the wall cavities of your house, with predictably disastrous results for the structure. A good and simple barrier is provided by stapling polyethylene sheets over the studs before you put up lath or sheetrock, and this is what we used. If you are lucky enough to have plaster walls in good enough shape to save, however, you can still provide a reasonably effective vapor barrier without tearing them down. There are vari-

ous paints formulated specifically to seal walls against water vapor, and although they are often more difficult to apply than ordinary primers, you should definitely take the extra trouble. This is one feature that older construction almost never provided, but which modern insulation and living temperatures have made a necessity.

Once we had taken care of our insulation and vapor barrier, the next step was to finish the walls and ceilings. Plasterwork is pretty much the same today as it's always been, except for a few changes in materials. Cellulose has replaced horsehair as a reinforcing agent, for example, and the lath, or backing onto

199

Installing insulation in an old building becomes more difficult when you don't want to disturb interior finishes like these handsome wood panelled walls at Guernsey Hall in New Jersey, or where you don't want to re-finish the interior surfaces at all. The designer of this converted barn in Connecticut chose to leave the original barn siding exposed, so any insulation must be laid on outside under a new exterior finish, or added behind the few new wood panels inside.

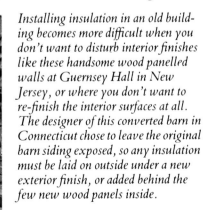

which plaster is applied, is no longer made of wooden slats as it was when the Bigelow House was built. Wood lath was first superceded by heavy metal mesh, which is expensive but much easier to put up. More recently, inexpensive gypsum panels known as rocklath or blueboard have emerged as the most practical lath for flat surfaces. These can be nailed up quickly on walls ▼ and ceilings in four by eight foot sheets. For ceiling work, as you can imagine, a couple of people are needed to hold each sheet up while someone else nails it in place. Special ridged nails are needed to support the weight, and when a particularly strong bond is needed, narrow Phillips-head screws driven by a variable-speed electric drill can be used.

Because blueboard is essentially a prefabricated three-eighths-inch layer of plaster, it simplifies the plasterer's job. Over other kinds of lath, three coats of plaster are traditionally necessary: a *scratch* coat first, a thick *brown* coat second, and finally a *white* coat to make a smooth, hard finish. Over blueboard, however, the *brown* coat can be omitted. Unlike the other kinds of lath, however, blueboard does not conform well to irregularities in the studs to which it is being nailed, so a certain amount of time must be

spent cutting and shimming most old walls to form an even plane. Where a wall or ceiling actually curves, as often happens in a stair hall, metal lath is far preferable to blueboard since it can conform easily to the curves.▼

Plastering looks easy when you see someone good doing it, but then so does playing tournament tennis. Gravity being what it is, a beginner can find it difficult even to get wet plaster from the trowel onto the wall or ceiling. Once there, a lot of skill is needed to ▶ smooth it into an even surface before it hardens. There aren't many skilled plasterers left, and we were very proud of the elegant,

uniform walls and ceilings which Sal Vassallo and his crew created for us.

On the third floor ceilings of the Main House, Sal produced a different effect than the flat surfaces elsewhere. Right after he had applied the final white coat, he went over its smooth surface with a wet brush.

The bristles gave the plaster a kind of grain, and the moisture drew sand down to the surface, making an interesting rough texture. This was quite attractive on the ceiling, but we wouldn't have wanted a sandpaper effect on the walls where you could touch it.

We made all the walls and ceilings ▼ at the Bigelow House of plaster, first because it is strong, durable, and uniform when properly applied, and second because we knew from our experience in Dorchester that Sal is an excellent craftsman. If you can't locate a really good plasterer, however, or if your budget can't pay for one, there are other alternatives. The most common substitute for plaster is *sheetrock*, a gypsum panel similar to blueboard, but with a smooth finish surface that can take paint. Instead of putting an entire coat of plaster over sheetrock (as you do over blueboard), only the seams between sheets are covered, using *spackle* (joint compound) instead of plaster. Whereas plaster hardens fairly rapidly while still damp, joint compound hardens very gradually as it dries. Although this means that it shrinks a good deal more than plaster does, it is very easy to work, both while damp or hard. Cloth or paper tape is used with the spackle to bridge the joints, over which several layers of spackle are applied, allowed to dry,

and sanded smooth. If carefully done, it is impossible to spot the joints on a finished, painted sheetrock wall, but few spackling jobs are that good, unfortunately. Nonetheless, sheetrock can provide an attractive and inexpensive interior wall and ceiling surface.

You may be surprised to discover when you examine your old house carefully that what you took to be an ornate plaster ceiling is actually pressed tin. Metal ceilings usually hold up very well, but they can be hard to repair. Fortunately, some patterns are being reproduced by firms that cater to the rehab trade, so it is not as hard as it used to be to replace damaged or missing sections of metal ceilings.

If you do have an original plaster ceiling with elaborate stucco ornamentation that you're determined to save, it *is* possible to mend sags, cracks and holes. Sagging occurs either when the plaster breaks free of the lath or when the lath pulls away from the studs. The latter problem can be corrected with additional nails or screws, which can then easily be concealed with spackle. But when the plaster breaks away, things are more complicated, and you may be best off consulting a conservator. One solution is sometimes to drill small holes in the plaster, inject glue into the cavity, and shore the loose cast back up against the lath until the glue has set.

Clearly, this can be tricky. Cracks and holes, on the other hand, you can easily fix yourself with spackle (for small ones) or patching plaster (for larger ones), followed by sanding and painting.

Even if you fall in love with the brilliant chalk white of unpainted plaster surfaces, ▶ you should still paint them; you can keep the color, but you should take advantage of the protection paint gives. ▼ Although a layer of paint is extremely thin, it is strong, resistant to abrasion, and waterproof. At the Bigelow House we took the advice of our painting contractor and chose oil-based paints for all the interior walls and wood work as well as for the exterior trim. In these days of latex, it is somewhat unusual to use oil-based paint on interior walls, but it still has distinct advantages. Not only does it tend to be more durable than latex, but its coverage is better, and it acts to some degree as a vapor barrier, which latex does not. Latex, of course, has the advantage of being very easy to apply and clean up.

All paint is made up of tiny pigment particles suspended in a bonding agent diluted with a volatile liquid base. Pigment is carried evenly over a surface by the base, which then evaporates, leaving the pigment sealed in the bonding agent. *Lead paint* is so named because of its lead carbonate pigment (a po-

tent poison) usually suspended in a base of linseed oil. Modern oil-base paint is made with a safer basic pigment, titanium oxide, in an alkyd resin base that is cheaper and more effective than linseed oil. Latex paint uses water as the volatile base and resin as the binder. It coats a surface easily, but doesn't tend to penetrate as well as oil-based paint, although it can bond *too* well to old paint, sometimes shrinking and pulling the paint beneath it off the wall.

One important characteristic of latex is its permeability, at some times an advantage and at others a disadvantage. Although latex is waterproof, water vapor can pass through it, making it an ideal paint for exterior siding. The blisters you sometimes see in old paint on an exterior are there because water vapor passing outward through the wooden siding couldn't escape through the oil-based paint, and formed little pockets. With latex paint this would not have happened, since the paint can breathe. But an oil-based paint is preferable on interior walls, which should be as impermeable as possible.

All our interior wall surfaces were new, and all the doors and woodwork had been stripped of paint at a plant, so we didn't have to bother with the tedious job of scraping and sanding before we painted. We used a satin finish enamel on the plaster, and a

semi-gloss enamel on the woodwork. A semi-gloss finish is easy to clean, and thus a good idea for doors, door and window frames, switch plates and any other surfaces that get a lot of fingering. Our decision to paint the old woodwork was made partly because it had been painted originally, and partly because we had to use new wood in patching damaged or missing pieces, so we couldn't have achieved a uniform natural finish otherwise. If it hadn't been for this, we might have been tempted to leave it unpainted, because the old pine was a beautiful honey color.

If you have a room of panels or moldings you're considering stripping, take into consideration the color of the wood. If it is naturally very dark, leaving it unpainted will often make the room feel smaller and more enclosed. Any woodwork left bare or painted a different color from the walls stands out sharply and may distract attention from features you'd rather emphasize. One way to put things in a better perspective is to paint the woodwork the same color as the walls but in a high gloss finish instead of satin. The shiny effect makes the woodwork impressive visually without overwhelming the room architecturally.

The treatment of stair finishes, if your stairs are wood, should be consistent with

206

The Stewart Residence in San Francisco has been saved and re-used by relocation and restoration. It was "shoehorned" into a very narrow building lot before the exterior was restored.

that of the other woodwork. This is really an exercise in logic; elements that serve similar functions should generally be handled the same way. For example, both the balusters and stair risers are vertical supports, so if you paint one you should probably paint the other. Similarly, if you leave wood paneling in a room unpainted, it would be consistent not to paint decorative carving or panels on a staircase in the same area. Stair treads are really raised pieces of the floor, if you think about it, so you might plan to stain them the same color as a wooden floor or carpet them in the same material. Of course, if the treads are made of a different material than the

floors (metal *vs.* wood or wood *vs.* stone), then there is no need to match finishes.

In most old houses the floors are wood, and often a few repairs and sanding will restore an old floor to its original splendor. We expected that to be true in the Main House, but we ran into a few surprises. Our first problem was structural: our flooring contractor informed us that the floors were weakened by having been laid in the same direction as the subflooring instead of perpendicular to it. We resigned ourselves to ripping them up and re-laying the boards.◄ Well, as any builder can tell you, removing old floorboards intact is extremely difficult. You can't count on saving all the pieces, and we lost quite a few to the crowbars. As we took them up it became clear why so many of them snapped. Our floorboards, like modern ones, were *tongue and groove*: each board has a one-quarter-inch groove along one edge and a corresponding one-quarter-inch tongue along the other. In the process of laying, the groove of a board is first fitted snugly onto the tongue of the previous board and is nailed in place through its own tongue. The next board conceals the nails. Our floors had suffered so much wear and been sanded down so many times that the boards were worn to within a sixteenth of an inch of the tongues, and so split easily when

207

we pried them up. It was clear that the ones that didn't split would rapidly wear out —if they didn't split when we tried to put them back down. Reluctantly we agreed that we needed new floors.

The new oak floorboards were the same width as the old ones (a standard three inches), so the floors we put down look the same as the originals. They were laid amazingly fast. Our contractor demonstrated for us how the nails can be both hammered in and counter-sunk in a single operation using a power nailing machine, cutting nailing time by seventy-five percent. ▼ Because we dispensed with borders which would have

208

required mitred corners, the floors went down in a matter of hours! The fancier parquet which we chose for the formal entrance hall in the Main House took a little longer, but was also easy to install. The stair treads were in good shape, so all we had to do was sand them and stain them the same color as the parquet.

The only limitation on your choice of stain for a new or newly-sanded floor is to avoid exterior stains. These are usually oil-based and designed not to dry out for a long time so as to protect the exterior wood. This means that a polyurethane or varnish coat over them may not adhere unless you have dried them first for weeks and weeks, and sometimes not even then. Polyurethane is a truly miraculous finish coat for floors, as it is easy to apply, tough, and does not need waxing. High gloss polyurethane is too slick and shiny for most houses; you'll probably be happier with a semi-gloss finish, which is about as shiny as a waxed floor.

Many old buildings have floors which never were meant to be polished or admired, and you'll find, as we did, that sometimes nice old floors are too decrepit to restore. It is possible to treat the old surface as sub-flooring and cover it with another kind of finish, such as carpet, vinyl or tile.

Vinyl is a very practical material for kitchens, bathrooms, utility rooms, or even children's rooms. It is waterproof, sanitary, easy to clean, and very durable. Vinyl comes in two forms, sheeting or tiles. The sheeting gives a single, uniform floor surface with few seams, and can even be coved up the wall to eliminate hard-to-clean corners. The sheeting is fairly thin and flexible, however, so it will reveal any cracks or imperfections in the surface of the subfloor, whereas tile vinyl is usually heavy enough to hide any irregularities. Vinyl is a great improvement over linoleum, in durability as well as variety of patterns and colors. There are even some manufacturers who will match any color sample you send them—at a price, of course.

Wall-to-wall carpeting is also a practical and comfortable solution to a rough floor. If the area involved is going to get a lot of use, a fairly tight, flat carpet with a dense rather than a deep pile usually wears the best and is easiest to keep clean. Industrial carpeting in a neutral color is a good choice. A deep pile is hard to take care of, and a large pattern or very bright color will claim a room for its own.

At the Bigelow House, carpeting made sense in a number of places. On the third floor of the Main House, for example, the original floors were rough planks. By cover-

209

Restoration of the facade of the historic Dullnig Building in San Antonio, Texas, is a good illustration of what paint can and can't do for you. The brick building had been painted, stuccoed over, and painted again before it was purchased and restored. The stucco was removed, the bricks repaired, and existing cornices and moldings *were repainted a contrasting color. A final touch was the addition of a painted, trompe-l'oeil cornice around the top of the building, as a substitute for a real cornice which could not be fitted into the budget.*

ing them with carpet, we gave the area a snug and finished appearance. Similarly, in the second floor of all the other units, carpet seemed to us the most pleasant and practical covering for the new floors.

Downstairs in the Barn and Stable Units, we finally had to give up on re-using the solid old planks which we had rescued from the original structure. Wonderful as they looked, they had been imbued with too many years of animal life to be sanitary for human habitation. A lot of sanding and bleach might recover such boards for your own house, but we didn't want to take any chances either with the Board of Health, or on behalf of the future owners. Still, it was hard to let go of the barn image, and we compromised by specifying new six-inch wide floorboards, which are at least somewhat reminiscent of the original planks.

The materials and finishes we've been discussing so far will account for most of the surface area of your interior. What remains are the surfaces which will take the most intensive and specialized use, and which will therefore require the most durable, impervious coverings. Floors, for example, in a kitchen, bathroom, or greenhouse area should be particularly resistant to water and dirt. Vinyl is the most economical material with these characteristics, but over small

areas you may want to consider ceramic tiles. *Quarry* tile is the simplest of these, made of unglazed fired clay with a flat ground finish. Not only is quarry tile resistant to water and very durable, it comes in a number of different colors and can be extremely attractive. We used it to cover the floor of the greenhouse in the Barn Unit and were very happy with the result. For kitchen floors, however, we felt that quarry tile would be a little too porous and susceptible to staining, so we chose a glazed terracotta which we hoped would be very easy to clean.▼

A dense, impervious ceramic tile is also

appropriate for bathroom floors. An enormous selection of shapes, sizes and colors of ceramic flooring is available, and one of the wonderful features of the material is the opportunity it offers to create a mosaic pattern using more than one shape or color on a floor. The smooth glaze of a wall tile can be dangerous on the floor because it becomes slippery when wet. In the master bedroom in the Main House, our expert tile-layer, Charlie English, produced a traditional octagon and dot pattern in black and white which looks restrained and exceedingly elegant.◄

In a bathroom, parts of the wall as well as the floor have to be waterproof. In fact, ceramic tile has no competition as a material for covering bathroom walls. Other than a single-piece fiberglass tub surround, or shower stall, it is far and away the best material for the job. The variety of designs available for wall tiles is many times greater than for floor tiles. You can pay as much as a hundred dollars a square foot for lovely handpainted replicas of historic tiles, or even have original murals painted in glaze on your bathroom walls. There are also plenty of relatively inexpensive and beautiful colors and designs to choose from.

Tile can be applied in several different ways. The simplest is to glue it up over sheetrock. The sheetrock which you would use in a bathroom is water-resistant, which usually has a different color paper finish and is slightly more expensive than the ordinary kind. A difficult and possibly stronger technique, the traditional *mud job*, involves mounting the tiles on a heavy coat of mortar over metal lath. Getting the mortar onto the lath isn't hard, but real skill is needed to smooth it into an even surface, and to stick the tiles to it without deforming it, getting them all straight.▼You'll be unusually lucky if the tiles perfectly fill the space you have to put them in; it is almost always necessary to cut the last one in each row. If you don't

need to trim much, tile clippers will bite off small chunks in a relatively straight line, but the neatest approach is to use a table saw with a carborundum blade, because it leaves an absolutely smooth edge.

Wall tiles can be used in places other than the bathroom. Particularly attractive places are around fireplaces and stoves. The original fireplace surrounds at the Bigelow House had been faced with fairly small tiles, about two inches square, with a deep, translucent glaze. We loved the effect, but the living room fireplace was missing many of its tiles. At first we were afraid that there was no hope of matching them, because modern

212

manufacturers don't produce anything similar. Then, in response to a plea to our viewers, Marje Paulsen,▼ a local potter, volunteered to reproduce them for us. The knowledge, skill and extensive experimentation she devoted to this project produced tiles beyond our wildest hopes—it was hard to tell the reproductions from the originals!

In the Barn Unit, we turned to an even more traditional material for the fireplace surround—stone. We had hoped to be able to find some local *puddlestone*, a rock which exists on our site, but we didn't think anyone quarried it any more. We resorted instead to a similar *fossil-stone* from New York

State, which has the added appeal of tiny petrified prehistoric creatures on its surface. It came to us in large chunks, some of which our stonemason, Paul Panetta, had to break up into smaller ones to make them manageable. Part of a stonemason's art is selecting and arranging the stones in a wall so that they fit together properly from a structural and aesthetic point of view. ◀ Paul laid them against the cinderblock chimney wall using very deep pointing and a dark mortar, so that from a few feet back the mortar joints are invisible and the wall looks almost as it if were dry laid. Metal tabs set into the mortar of both walls tie the fossil-stone to the cinderblocks, preventing the surfaces from separating. Over the hearth a beutiful bluestone slab was set into the wall as a mantle. The overall effect is massive, almost monumental—truly a fireplace around which the whole family can gather on cold winter evenings!

The most expensive surfaces (per square foot) in your house, and some of the most useful, will probably be countertops. Counters can be covered in any number of materials—plastic laminates, ceramic tile, stainless steel, granite, marble, slate, soapstone, butcher block, Corian (a DuPont plastic which looks like marble but which is practically indestructible), and many more.

Plastic laminate seems to be the automatic choice of budget-minded builders, but it is actually not so very much cheaper than other materials, especially when you consider that it doesn't last as long. Stainless steel is certainly durable, but many people are put off by its cost and by its lustrous, industrial appearance.

Among stones, granite is probably the strongest, but because it has to be quarried in slabs no thinner than about two inches, it is very heavy. It is expensive to buy and it costs a lot to ship and install. Marble, by comparison, is easy to work with but is so delicate that it functions well *only* as a pastry surface. Most people don't realize how

216

easily it can be stained by oils and pitted by fruit juices, vegetables, alcohol, or sugars. There is no prettier stone, but be cautious about employing marble as a functional surface. Slate and soapstone do make durable, if less glamorous, surfaces, and they weigh less than granite.

Butcher block is a popular surface which stands up well, but it too has a drawback. As it gets older tiny crevices in the wood tend to fill with food particles and become homes for thriving colonies of bacteria. In order to keep a thoroughly sanitary surface on butcher block, you must scrub it roughly with a heavy wire brush or sandpaper as you may have seen your butcher do.

In many ways, Corian seems to be an ideal compromise between these various materials. Its only weakness is that it can be scorched by a very hot pan. Although it's harder to scratch than a laminate, it will, like almost any surface, show small scratches and scoring if you chop on it, but because it is a homogeneous material you can gently sand it down to remove any really disfiguring gouges.

A wider variety of colors at a lower price than any of the materials we've been discussing is still offered by plastic laminates. They are very easy to keep clean, and as long as they're not mistreated, should last for many, many years. Installation can be a problem, though, especially if you've never done it before. The material comes in four feet by eight feet sheets that are about one-sixteenth of an inch thick, and must first be cut into pieces a little larger than the surface to be covered. A fine saw or scoring knife will cut the laminate, but it is somewhat fragile at this stage, so you must be careful. You glue it in position with contact cement, which means that alignment is the key to success. Once the plastic touches the counter, it sticks tight—forever! If it happens to be an eighth of an inch too far to the right, that's where it stays. One way to get around this difficulty is to glue the whole sheet to a sheet of plywood first, and then cut the countertop out of the resulting union. ◀ Still another trick is to separate the two pieces to be glued with dowels laid at regular intervals between them. When they are perfectly aligned, the dowels can be withdrawn slowly, one by one, allowing the pieces to come together. No matter how you do it, the process takes practice, and learning can be messy and expensive.

217

Interior finishes in re-used buildings can demand as much work as this elaborate carved and painted ceiling restored at Guernsey Hall, or as little as the completely unfinished wall surface in a barn which has been made into a summer home.

Chapter 9

Fixtures and Features

By now we had a pretty good idea of what our units were going to look like. The spaces were clearly defined; new windows and skylights were in place, so we could see the effect of natural light; and we could tell from the various tiled, carpeted and wooden surfaces what the atmosphere would be. Even before we'd finished shingling and plastering we'd begun to turn our attention to the next level of refinement—the architectural details, hardware, fixtures, appliances and equipment that set the final tone of a house. If finish work can be compared to icing on the cake, then these fixtures correspond to the candles and decoration.

The job of choosing all the little items to complete the renovation sounds insignificant compared to what we'd already been through, but we learned in the Dorchester house not to underestimate what a difference these decisions can make. The original details of an old building give it its individual character and the details you add in the finish stages of rehabbing will invariably shape this personality. Because even the smallest decision of this sort affects the final appearance, the selection of fixtures and features should be given your full attention. It is, of course, much easier to make choices for your personal use than for a hypothetical buyer. What may seem appropriate and

Not every old building has as distinctive existing features as these jail cells. A restaurant chain which makes a practice of converting interesting buildings for their restaurants has transformed this old jail in Concord, New Hampshire, into an unusual steak house.

beautiful to you may leave someone else cold, or even offend them. And, believe it or not, some prospective buyers will reject a house or apartment just on the basis of cosmetics; in fact, I've done it myself. You can't solve the problem simply by buying the *best*, because where taste is concerned, there is no best. The best you can do is to be consistent.

Of the many different kinds of features and fixtures you'll have to consider, the most costly have to do with finish carpentry. The most numerous will be the visible elements of the mechanical systems. Few homeowners have any idea of the variety and range of products available. Because you rarely find a complete selection in any one place, it takes a lot of determined research to get to know the market. Architectural libraries and offices usually keep a fairly comprehensive file of product literature, and there are a few catalogs and periodicals that advertise architectural products, but no one source comes close to showing it all. If you select the details yourself, you will have a fascinating but time-consuming period of searching, thinking, coordinating, eliminating and inventing. It's the one activity rehabbers find most addictive.

At the Bigelow House, we were racing the clock by the time we reached the finish work because of all the unforeseen foundation work. We didn't have time to explore thousands of catalogs or make pilgrimages to architectural salvage yards as we would have liked. Instead, we decided to use good, fairly standard modern design for all items that we did not restore to their original form. There was not a great deal to restore, because modern habitable interiors had never existed in any of the units except the Main House, and because vandalism and decay had obliterated many of the fine details there. The conception of spaces was so modern in feeling in the other units, especially the Barn, that we were not even tempted to choose historical details. In all new work, however, we did try to equal the quality and elegance of the original construction. We wanted to make sure the visible details were as good as all our hidden work. Keeping this in mind, we often found ourselves selecting hand-made rather than high-tech materials and hardware.

The restoration we did do at the Bigelow House was by no means easy. Stair-rails, banisters, balusters, paneling, molding, doors, fireplace surrounds and hardware in the Main House Unit all took a lot of attention. As is usually the case, the process of restoration posed all kinds of problems no normal contractor ever encounters. Often when we ran into trouble, a plea to our

The double helical stone staircase at the chateau de Chambord in France is one of those architectural features which is probably more famous than the building containing it.

viewers placed remarkable amateur talents at our disposal. Of course, as a private rehabber, you don't have a television audience to solicit, but you *can* put a classified ad in the paper, and before you do that, you should ask around among your friends and neighbors. You'd be surprised at the skills and know-how hidden in the most unexpected places.

Another way to replace lost ornamentation is to explore architectural salvage yards. A good one may be worth a trip of several hundred miles because they often carry irreplaceable finds, but you may not have to go so far. Some contractors specializing in restoration work keep their own stockpile of architectural details saved from different jobs or purchased from demolition crews. Another possibility is to shop through catalogs of the various millwork companies that reproduce historical woodwork—quite a few of these have sprung up in response to the growing popularity of rehabbing. In our case, however, we needed to look no farther than our viewers for hand-milled replicas of the balusters and handrail for the Main House stairs.

Staircases
Down through history, architects have been fascinated by the infinite opportunities for creating spatial complexity and visual delight with stairs. Many famous stairways are better known than the buildings that house them. In Victorian architecture, staircases were especially prominent and lavishly decorated, and Richardson was well known for his ability in this area.

Not only do stairs lend spatial drama to a main entrance, they can also serve a host of specific needs. The placement and design of fire escapes and enclosed fire stairs, for example, is determined by the building code's requirement that every apartment have fireproof access to the street. Many types of folding stairs have been designed to take up a minimum of space, wheel-chair users need ramped stairs, and crowds can best be accommodated by moving stairs. An impressive formal entry can be made on a massive flight of marble and onyx with richly carved and ornamented rails, while one person can ascend or descend on a narrow spiral of wood or steel, arcing out and down from great heights with no visible means of support. It is a pity that people have lost some of their interest in stairs with the advent of skyscrapers and elevators, because there are as many beautiful staircases as there are houses.

The first time we saw the formal, traditional staircase in the large entrance hall of the Main House, even with most of its bal-

221

At Guernsey Hall, all the residents of the new apartments enjoy use of the grand stair which was once a central decorative feature of the large house.

The magnificent carved staircase at the Ames-Webster House in Boston is part of the corporate entrance hall— hardly a "lobby"—for the architectural, legal, publishing and planning offices which now inhabit the former urban palace.

usters missing and only a suggestion of the original woodwork left, we felt at once that this was one of the grand features of the house and deserved to be restored to its original state.▼ The run of stairs from the second to third floor was not nearly so large as the first flight, but with its graceful banister and curved base moulding, it also added a lot to the character of the house. ▶ It was a challenge to live up to this standard in the new stairs we had to provide for the other four units.

The main stair we installed in the Barn Unit also rises from an entry hall, but its effect is very different from the traditional formality of the Main House. Starting on your left as you walk in the front door, it curves up in a tight, smooth *U*. Above and ahead you can see over the second-floor landing all the way up to the new skylight on the opposite side of the roof peak, giving you a sense of the Barn's original scale and openness. Jock managed here to make the stairs dramatic without taking up too much

room, and at the same time provided them with natural light. He used the same device on a smaller scale in the Icehouse Unit; the entry is lit by a skylight directly above, and you can see up over the landing to the skylights of the study.

In the Woodshed Unit, when you come in the front door, you are immediately struck by the large living room with its tall cathedral ceiling. As you stand admiring it, your attention is drawn to a mysterious balcony jutting out into the air on the left—the landing of the stairs to the second floor! Every time the owners pass up or down, they will enjoy the double-height volume of their living room from mid-air.

Another interesting twist is seen in the secondary stair in the Barn Unit, a compact spiral supported on a central post joining the study upstairs and the living room. A glance at the plans will indicate that stairs of this sort take up much less room than straight or rectangular ones. Except in very large sizes, they are impractical as a main stair because of the difficulties they present to furniture movers (the smallest models, with overall diameters of as little as three and a half feet, are a tight squeeze even for one person), but as secondary stairs, their sculptural forms can be very beautiful. We were excited about the effect of the graceful wooden spi-

ral in the tall open space under the large western gable.

Custom-made spiral or helical stairs can be very expensive, especially if they are self-supporting instead of being fixed to a central post or column. Free-standing stairs, of course, are visually more exciting than the central post type, but they are more difficult to build and almost always have to be custom made. There are many models of the central column type, on the other hand, which can be bought prefabricated in wood or metal quite inexpensively, some even in kit form. Both our space requirements and our budget made a central column spiral ▼

the natural choice for our situation, but if you have a large, double-height room and can afford it, a free-standing helical stair can be quite spectacular.

The only apartment in the Bigelow House with an attic or a basement is the Main House Unit. As is usually the case in an old house, the existing stairs were sound but very simple, and we were content to simply repair the plaster finishes on the walls. Entrance to the crawl spaces in other units required nothing more elaborate than trap doors in closet floors. If you do need to provide access to an attic, however, and don't want to devote space to a full stair, product catalogs will show some other possibilities that may prove sturdier and more attractive than the standard drop-down folding stair door commonly used by builders. Look into ship fittings for retracting or folding hatch stairs, and in library equipment catalogs for hanging, rolling, pivoting or collapsible stairs. There are some wonderful designs available, if you have the patience and ingenuity to unearth them.

Doors

Because of the obvious practical importance of entrances and exits, designers have always given them a lot of attention. Conventions of ornamentation have developed not only

224

The architect who converted this Long Island stable to a private home has managed to include a human-sized front door in the original carriage entrance and a large picture window in a gable overhead without disrupting the scale of the building.

for the doors themselves, but also for the frames around the openings. Today, doors are available in a vast array of materials, in every possible style, and moving in any way you wish (swinging, folding, sliding, or rising). In spite of the vast selection, it is usually impossible to find a perfect match for an old door, and you will have to make a choice among modern versions or have a new door custom made. Once again, keep the original architecture of your building in mind when you decide what you want, but don't be afraid to use your imagination.

Almost everyone is conscious of the design of a front door as the formal point of arrival, and we felt that planning the individual approaches to our units was one of the most important aspects of design in this project. The front entrance to the Main House is immediately apparent to anyone approaching the structure, but we had to find a way to make the other four units equally prominent, while at the same time not disturbing the original facade or distracting attention from the original entrance. This involved the placement of the doors more than it did their actual design. One solution was to locate the Barn Unit's front door on the north side of the building directly across from the garage, so that visitors arriving by car will be able to spot it easily,

and where it doesn't compete with the approach to the Main House. Since the three units between the Barn and Main House are gathered around the stableyard, it seemed natural to use it as a collective vestibule for their entrances. The old stableyard gate, which had originally closed with huge sliding doors that are still in place, provides a perfect formal entrance to the yard. Our plans called for elegant but simple landscaping around the courtyard to draw a visitor's attention into the area and lead the eye to the front doors of the Woodshed, Icehouse and Stable Units. With only three entrances in the courtyard, confusion is avoided and a certain degree of privacy is retained. We particularly wanted each unit to have a distinct entrance equal in convenience and formality to the others.

We used traditional wood paneled doors for each entrance, in keeping with the historical appearance of the exterior facade. The front door of the Main House, a 150-pound, two and five-eighths inches thick, solid pine beauty, was still in place when we bought the house. Vandals had damaged the side of it trying to pry it open, but it was still in pretty good shape. A little restoration work made it as good as new. We had to graft in a number of patches, especially where the old hardware had been removed,   225

In New York's Soho district, a creative loft dweller has embellished the riveted steel service door which is the entrance to his apartment with paint and a unique welded sculpture.

The monumental doorway at the Mergenthaler Linotype Building in Chicago was scaled down to residential proportions when the building was converted to apartments. A clear contrast of old and new materials emphasizes the new use.

but once they were filled and painted, you couldn't tell. Without duplicating the full weight of this door, we tried to reproduce the same style and solidity in the front doors of the other units.

If you are adapting a non-residential building to residential use, it is sometimes difficult to make the front door recognizable as the entrance to a home. Loft-dwellers often find themselves using an old service door as their main entrance, which is hard to distinguish from real service exits or even janitorial closets. It is usually an advantage for security purposes to retain these plain metal doors, but they can be painted or embellished

to become unique and instantly recognizable.

Most of the interior doors in the Main House were still hanging, although all the brass hardware had been stolen from them. Their thickness and elegance recalled a more opulent building era. ◄ These interior doors are as thick as standard *exterior* doors are today, ▼ and our old front door is another inch thicker than that! Since we had made only minor changes to the plan of the Main House, we had enough original doors for our purposes there. In the few places where we installed new closets, we decided to put in plain, flush wooden doors. There were not enough of them to be really incongruous, and it seemed appropriate to downplay the doors hiding new additions like the laundry machines.

Anyone who hasn't tried to build and hang a door tends to take them for granted. A door is a door—they open, they shut, sometimes they stick, and if need be you can take them off their hinges and use them as desk tops. If you've ever undertaken to reconstruct an old paneled door, however, you will realize that a lot of wood, time and know-how went into each one of them. If you've tried to hang one, especially in a new wall opening, you will know that getting a tight, level fit is a time-consuming, tricky

process. A door, after all, is really a mechanism, and as such is much more complex than the fixed wall surfaces surrounding it. That doors open and close so dependably is a tribute to their engineering, and if we find so many of them in old houses, it means that our predecessors were generous and industrious. In recent years it has been fashionable to discard doors in the process of renovation in order to open up the spaces in old houses. While there is nothing wrong with the aim, in practice, removing doors can often have little beneficial effect. Not only are they irreplaceable features of your old house, they also contribute to its energy efficiency and comfort by letting you divide it into separate temperature zones.

Windows and Skylights

We've talked a lot about the problem of finding windows that would fulfill the Historical Commission's requirement that we restore the original facade of the Bigelow House. Where we put new windows in the side walls, we used replicas of the double-hung Main House windows. ▶ All the old window openings were preserved, so each unit has at least some original detailing, but the repair and copy work is so good that it will be hard to tell the old from the new.

Not all the windows we added are tradi-

tional in size or shape, but in all cases we tried to locate them in ways that would not impinge on the original design. For example, Jock recessed a picture window in the opening of the old gable door, thus adapting the original opening without greatly changing its outside appearance. The greenhouse

is also a straightforward newcomer ▼ —a modern version of the original sunbath— but Jock carefully designed it to occupy the place of the old carriage doors so that it doesn't interrupt the exterior lines of the barn.

228

Windows are more complicated than doors, but because one looks through instead of at them, they tend to be simpler aesthetically. They vary mainly in size and proportion, and you will have relatively few decisions to make about style. If you choose traditional double-hung windows, you should have them custom-made by a reputable window company. Although it is slightly less expensive to buy standard sizes, it is well worth the small added expense to get the exact dimensions and mullion configuration your house requires. If, on the other hand, you are installing new sliding or picture windows, you should try wherever possible to use sizes that are standard for the type of window you want.

With energy prices going up all the time, you will almost certainly want to ensure that your windows have more than one layer of glass. Storm windows, although they don't look authentic on a Victorian facade, are the most practical solution for old houses. Where you are installing new windows, however, you should consider double- or triple-glazed units. Windows can be a major source of heat loss, and studies suggest that in most climates double-glazing soon pays for itself. Good windows will also keep you from air-conditioning the garden.

Skylights are practical and dramatic additions to many old houses. Again, there is a far greater variety of them available than preliminary research would suggest. You can buy multi-pyramid designs, polygonal domes, ridges, panels and tunnels, many of them mounted so they open. Keep in mind, however, that all skylights tend to leak and should therefore be carefully installed by an expert, with plenty of flashing, preferably near the peak of a roof.

The best models come with flashing built into their frames and a drip-guard to catch run-off from condensation, which can be a nuisance in humid areas like bathrooms and kitchens. Double-glazed skylights are much less likely to form condensation than plastic ones, and because of their insulation value they lose much less energy to the outdoors. If you don't insist on a clear skylight, you might consider a model constructed of insulated, translucent glass blocks. Fiberglass filling makes these panels almost as heat-tight as the surrounding roof.

At the Bigelow House, we installed very handsome, flat panel skylights with their own flashing in an extremely well-made frame. The outside of the frame is metal, and can be painted to match the wood trim of the house, while the inside is made of a beautiful light wood, which looks wonderful in the upstairs rooms of our new units. These sky-

lights open from the top, so that you never have to lean out over the roof. In some places they're too high to reach without a pole, but their large handles are easy to hook. The ventilation they will provide should make the upper story considerably more comfortable in warm weather.

Moldings and Ornamental Woodwork

One of the joys of an earlier age was the elaborate and elegant use of wood. Although some paneling and carving is strictly decorative, the moldings, ▶ however ornate they may be, do have a function. Most construction requires a finishing cover to mask the uneven joints where walls meet floors, window frames, door jambs, and sometimes ceilings. The jaggedness of a joint depends on the construction techniques, but in general, any seam where two different materials meet should be covered with molding. In addition, chair-rails, picture moldings, corner pieces and baseboards serve to protect the walls against daily wear and tear.

We tried to be very careful when we removed the original wood moldings and baseboards from rooms in the Main House before we began demolition, but there were inevitable losses. By the time we reached the point of putting in window casings and door frames, ▶ we realized that we didn't have

230

enough of the old woodwork to do the job, and in order to complete the installation we had the profiles copied by a local millwork yard. It turned out that new moldings were easier to install than the originals; we had had to rebuild the windows and the old sections no longer fit together properly. Although we were able to use most of them anyway, you can tell that our finish carpenter was working with new wood when you see a perfectly mitred corner in the door or window frames. Once the moldings and baseboards went in, the rooms really began to look Victorian again, and the triple-hung reproduction sashes finally seemed to belong in the house. We saw that our care in getting the mullions and pane divisions right had been justified; all the details work together in a single, consistent, harmonious style.

Not only can you save and refinish the woodwork you find in your old house, you can sometimes introduce appropriate details salvaged from other houses of the same period. Although it takes patience and know-how to locate the right elements in salvage yards, you can restore a house much ravaged by later renovations to the splendor of the original period. Some people get so hooked on salvage yards, however, that they end up living in a hodge-podge of paneling and or-

namental woodwork of every sort. There is nothing to stop you, of course, from installing a courtroom wall, a bit of pulpit, a banister from an ocean liner and six feet of eighteenth century baseboard in one house, but you'll probably be happier if you respect the particular character of the house you're rehabbing.

Cabinetwork

Although cabinets and cupboards can be selected at the last moment, it is always preferable to start thinking about them at the very beginning, so that you can leave the right amount of space for the units you like. Once the dimensions have been settled on, finish decisions are less urgent. If you have selected prefabricated cabinets, however, they should be ordered in plenty of time so they can be installed as soon as the space is ready for them. Often when you are dealing with an old house, the room dimensions are predetermined and irregular, making custom cabinetry necessary. Built-in features such as bookshelves, window seats, pantry cupboards, or even ornamental china cabinets are beautiful examples of a carpenter's talents, but generally quite costly. In the kitchen and bath it is more economical to have cabinets made up in the dimensions you require by a kitchen cabinet subcontrac-

231

New closets built around an old set of closed cupboards and drawers adds practical storage space in an old house and makes a handsome wall dividing two rooms.

Restoring old paneling in a new location and building reproduction cabinetry in historical detail is not the cheapest way to finish a wall, but it can be a showpiece of carpentry skills.

tor, or to choose among the wide variety of standard prefabricated cabinets on the market.

At the Bigelow House we had solid oak cabinet units made up for the kitchens of the Main House and Barn Units. These are substantial, even luxurious cooking areas, and we felt that they deserved fitting cabinets. In the limited space of the Stable Unit, we tried to design an efficient and attractive kitchen using cabinets of high quality, but without having anything specially made. It was a challenge, because we had to sacrifice some of the cabinet space to the hot-water heater, but with a little time and effort we were able

to achieve an elegance as compact as a well-fitted ship.

Deciding on kitchen styling must be the most subjective choice made when rehabbing; after all, no two people would furnish a room in the same way, and in the kitchen, most of the furniture has to be built in. Choosing for someone else, of course, is very hard, all the more because here you can't use the excuse of adherence to original details. Modern technology has so changed kitchens that they are best treated as completely new additions to old houses. Of course, traditional materials like tile, wood and stone can make a modern kitchen compatible in feeling with an earlier period, but the appliances remain distinctly twentieth century. For this reason, you are really free to choose any style that appeals to you, as long as it is reasonably consistent (an ultramodern range set in baroque cabinets will look bizarre). We finally settled on a restrained neo-colonial style right in the mainstream of American taste. Choosing fixtures is less difficult because they are primarily functional, but you will find them in more configurations than you can imagine.

Kitchen Equipment

The gadgetry available to kitchen planners is staggering in scope. Manufacturers really do

New kitchen design in old houses can vary from historically sensitive to frankly modern. A kitchen in a restored colonial house has been planned with a central work island containing a stainless steel sink and dishwasher, but the cabinetry is all hand-made cherry cup-boards in a colonial pattern. In contrast, the modern kitchen is all white formica and stainless steel.

enable every dream to come true. You can buy a whole kitchen, complete with refrigerator, counter, range, oven, sink and storage, in one piece measuring only six feet across. These units are useful in tiny apartments or as auxiliaries, but they *are* doll-sized, and not convenient for cooking much more than breakfast. Thinking bigger, the basic appliances a builder installs in most kitchens are a range, range-hood, oven, sink, dishwasher, and sometimes a refrigerator. Beyond that, there are indoor self-cleaning barbecues, microwave ovens, trash compactors, garbage disposals, built-in blenders, and much, much more.

Ranges can be gas or electric and can be purchased as countertop units or in combination with an oven. Many cooks prefer gas ranges both because the heat level can be changed instantaneously and because you can tell how much heat you're cooking with just by looking at the flame. If you decide on a gas range, it is a good idea to buy one lit by an electric spark rather than by a pilot light. Omitting the pilot light eliminates three problems: you don't have to worry about cleaning and adjusting it, you don't have to worry that it will go out when you're away and leak gas into the room, and you don't have to pay for a small but steady stream of

gas to feed it. Electric ranges have their enthusiasts too, especially the smooth-top ceramic units available today. There are also some extraordinary, futuristic European designs on the market. There are modular systems which let you put together any combination of gas and electric burners you like in a great variety of layouts, all with beautiful twenty-first century details. Parts and service, of course, can be a problem in this country with an imported product, so you should be sure to check that out carefully before you make a decision.

Ranges come with a number of extras beyond the basic heat source. Some models are available with interchangeable components, so that you can add a griddle, grill, or rotisserie. The range we installed in the Main House Unit had all three of these accessories, both because we wanted to demonstrate them on the air, and because our real-estate adviser had recommended that we make the kitchens and bathrooms as sophisticated as we could. The same range also has a very useful feature, a built-in venting system that sucks cooking fumes and grease into an exhaust vent at surface level, making a range hood unnecessary. When we grilled a couple of steaks on the indoor grill attachment to demonstrate the equipment for our viewers,

234

we were impressed that the system really did seem to eliminate the smoke. You still need a vent duct leading to the outdoors, and an exhaust fan, but it is both a convenience and a pleasure to be able to dispense with a hood over the stove. You can use the space you save to put up special lighting or a wall oven above the range without having to worry about grease buildup.

Those who aren't sold on a stove-top exhaust system have a variety of range hoods to choose from. If you do much frying or grilling, it is definitely worth getting a model that vents to the outside. The hoods with a filter and no vent are not very effective, and the filters have to be changed all the time if they're going to do any good at all. Of course, you'll have to change the filters on outside exhaust models too, but they really work, especially if you get a powerful fan. For more detailed information on range hoods and all other kitchen equipment you're considering, it is very helpful to consult the consumer guides that test this equipment.

If gas ranges are preferable to electric ones, the opposite is true when it comes to ovens. Electric ovens tend to heat much more evenly and dependably than gas ones. No matter which you choose, however, there are some extras you may want to con-sider. Insulation shouldn't really be an extra, but it is. Very few ovens are carefully insulated and sealed, which means that a lot more energy is required to keep them at a steady heat, and the temperature inside tends to fluctuate a lot. The self-cleaning capability is a convenient extra that eliminates one of the worst kitchen chores.

Beyond this, many ovens are available with sophisticated electronic controls. The model we installed in the Main House kitchen, for example, has a small memory that can be programmed to turn the oven on or off automatically at predetermined times and can make adjustments in the temperature settings as well. Appropriately enough, these computer-ovens are designed to look very futuristic, with black glass doors and digital controls, giving the impression that this oven leaves you no excuse for burning the roast, provided it doesn't de-materialize to Alpha Centauri first.

Because of its versatility and control, the kind of conventional oven we've been talking about is a basic tool of any kitchen. Relative novelties such as microwave and convection ovens can supplement it but not yet replace it. To begin with, microwave ovens are fairly small, do not accept conventional metal cooking vessels, and may conceivably pose health hazards. As for performance,

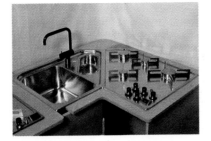

Modern manufacturers of counter top ranges have produced strikingly handsome designs and remarkable versatility in their equipment.

they are great for some jobs but make a mess of others. Food critic Craig Claiborne, for instance, reports complete failure with microwave soufflés. Roasts tend to come out looking gray and boiled instead of crispy and brown. Some foods, particularly those with skins, like potatoes, have a tendency to joyously burst. On the other hand, a microwave oven is fabulous for defrosting frozen food and reheating leftovers and has no equal for quickly cooking up a frozen dinner. Convection ovens are still the kind of novelty that microwaves were ten years ago. Most of the ones I've seen are fairly small, counter-top appliances into which a turkey, for example, would have a hard time fitting. However, new models are appearing which can supposedly accommodate an eighteen-pound bird. Convection ovens are essentially ordinary electric ovens with fans added to circulate the hot air rapidly around the food—not by any means a dangerous technology. Food does cook more quickly this way, and roasts brown and pie crusts turn golden. Many convection ovens have chambers of clear glass, and the cooking process can be fascinating to watch.

Having talked about everything but the kitchen sink, we should not neglect that important piece of equipment. Traditional enameled cast-iron has largely been replaced

236

today by stainless steel, which does not stain and looks very professional.▼ There are a few things to remember when you buy a stainless steel sink, however. First of all, be sure to get a reasonably thick grade of steel—many of the bargain sinks are very thin and don't hold up well under constant use. Also, don't think you have to settle for a standard shape or size—practically anything you can think of is available if you look for it hard enough (round, square, triangular, deep, flat, single, double, triple, you name it). For instance, you can sometimes save space by putting a triangular sink in the corner of a counter instead of using the standard rectan-

gular kind. If you want a less twentieth century effect, you can also consider soapstone, but it is quite expensive. An old-fashioned soapstone sink, although hard to find, heavy to transport, and tricky to install, is a pleasure to use, even in these high-tech days.

An extension of the kitchen sink is the dishwasher. This and a host of other appliances (refrigerator, garbage disposal, trash compactor, etc.) should be purchased on the basis of your common sense. ▶ Consumer guides can give you a good deal of information about particular brands and general standards of performance, but part of your decision must also be based on local service and pricing considerations. My only general rule is that if you go to the trouble of rehabbing a house, it's worth installing appliances that are as good as the rest of the construction.

Bathroom Fixtures

We Americans are a little odd about bathrooms. They are terribly important to us, as any real estate agent will tell you, but at the same time we almost always insist on packing them into tiny closets no larger than five feet by seven feet. One of the first things you might want to consider in planning for your master bathroom is a more European space —put it in a real room, with more than one

window. If you have enough space to do this, you will find you can create a completely different atmosphere. In a more public and recreational vein, you can also indulge in hot tubs, saunas, sun lamps, and even steam baths, but for most of us these sybaritic delights fall into the category of entertainment.

The basic distinguishing features of a bathroom are a sink, a toilet, and a tub and/or shower. Of these, the toilet offers the least field for choice. Most toilets are made of porcelain, though there are some lighter models made of plastic. And though they vary little in function, there are some options in design.

237

Both a simple, modern design and a simple antique version make excellent towel bars in old houses.

When it comes to bathtubs, your choice is a little broader. Cast-iron enameled tubs are still available and remain the most durable choice, but they are expensive to buy and costly to install because they are so heavy. Enameled steel is lighter and less expensive, but also less solid and durable than cast-iron. Fiberglass tubs are the least expensive, lightest, and least durable. Nonetheless, they hold up quite well as long as you don't use any abrasive cleaners on them, and they are much easier to repair when the finish is damaged. A great advantage of plastic is that it can be formed into a seamless one-piece enclosure for tub and shower walls, making the whole area handsome and easy to clean. Plastic is also made in a wider range of colors than enamel finishes, and the variety of shapes and sizes both of tubs and of one-piece shower stalls is far more varied than what you can find in metal. Not only can some of the designs help you save space in clever ways, but their striking looks can produce effects you could not achieve using more traditional forms. For example, if you're considering putting your bath in a real room, rather than being forced to cram the shower into a corner or above the tub, you can use a free-standing cylindrical shower stall made of clear or tinted plastic as the central focus of the space. Before you set

your heart on a plastic unit, however, check with your plumber—the building code in some localities won't permit them because they melt in the intense heat of a fire.

Except for the Main House master bath, we installed typical American bathrooms in the Bigelow House, using top-of-the-line equipment everywhere. Because we needed to conserve every possible square foot in those places, we really couldn't afford to make the bathrooms larger, much as we would have liked to. Our one concession to luxury was to use large oval whirlpool tubs wherever we could. In the existing master bathroom of the Main House Unit, we worried at first that a modern whirlpool might look out of place in the larger space with its Victorian details, but in fact the tub seemed particularly appropriate to the room's generous dimensions and pleasant orientation. In the Stable Unit, where tight economies of space forced us to forego a whirlpool, we hit upon surrounding a standard-size steel tub with Corian, the DuPont plastic that resembles marble. Even in cramped quarters, we felt that this would help make bathing the delight it deserves to be.

Corian has also made possible some exciting new developments in bathroom sinks. It is now possible to get an attractive Corian counter-top which dips smoothly at the cen-

238

An odd corner next to a window can be turned into a pleasant washstand. This old fixture is worth keeping for its location as well as its marble top.

Often, traditional designs in bathroom hardware are more attractive and easier to handle than those available from the big plumbing companies.

ter to form a durable sink. Not since the most luxurious days of stone-carving has it been possible to obtain a counter and sink made of one continuous piece of material. Traditional porcelain sinks come in a wide variety of shapes and colors and are available in beautiful, hand-decorated designs. When you are looking into modern sinks (and the same is true to a lesser degree for tubs), don't forget that an antique in good condition will function as well as a new sink and will fit in perfectly with the rest of your house. In fact, there may already be one installed in your bathroom! The main problem with old fixtures is that years of scouring may have worn away that shiny porcelain finish,

leaving a porous surface below that is easily stained and perhaps unhygenic. But, if you are lucky enough to find an elegant old sink which is not badly worn or chipped, you should use it.

When you come to select faucets and spigots, you can spend a lot of time looking at catalogs at plumbing-supply houses and shopping in custom fixture stores, marveling at the number of styles available. As you're trying to make up your mind, don't neglect to consider laboratory and industrial fittings, some of which are both practical and beautiful, and may serve you better than the standard household designs.

Hardware

Related to the faucets and spigots, which constitute outlets and controls for your plumbing-supply system, are a whole series of other necessary hardware items. Light fixtures, switches, plates, sockets, door stops, handles, hinges, locks, drawer pulls, coathooks, radiators, thermostats, fans, and even woodstoves are all items to be selected and ordered in advance if you want your house to look as you imagine it. Choosing what you want from among the thousands of designs available is best done by defining a single effect you're looking for, and then making individual selections within the range of acceptable appearance on the basis of performance and cost. We used modern design for all fixtures we added to the house because most of the interiors were new.

We tried to make the new lighting as unobtrusive and attractive as possible, steering clear of period-piece fixtures. Where light was needed in a specific location, we recessed lights into the ceiling or mounted modern fixtures so that they wouldn't be conspicuous. Recessed fixtures work best when they can perform a clearly defined task; otherwise a change of furniture arrangement would leave them useless. We used them to light kitchen worktables, built-in bookcases, fireplaces, and showers. In living rooms and bedrooms where furniture placement is arbitrary, we wired the electrical outlets to a switch by the door, and we expect that portable lamps placed by the future owners will satisfy other needs. We occasionally placed lights for pure dramatic effect, such as in the high open space above the stair hall in the Barn Unit, making the space come alive at night. We also put some outdoor lights in the garden to provide a night view. Only in very formal areas like the entry hall of the Main House did we resort to hanging fixtures.

The switches, plugs and switchplates we chose are modern in appearance. The switches are broader and flatter than the normal toggle, and the whole assembly of plate and switch has a neat, racy look. Dimmer switches (rheostats) in this line operate from the heat of your hand, like elevator buttons or time machine controls. The matching wall outlets are also nicely designed, and we felt that because of their simplicity, these fixtures would actually be less obtrusive than more conventional ones. Another way to downplay the appearance of electrical outlets is to install a baseboard strip that contains sockets at regular intervals along its length. If you place them carefully, they can often blend right into the base molding.

240

Spigots and faucets can be as simple and elegant as this one-piece fixture.

The sculptural quality of this antique sink does not seem at all out of character in a rustic bathroom, while sleek modern counters might.

The same principle of hiding fixtures at baseboard level also applies to radiators. Many models, whether electric resistance, hot water or steam, are sized to match old-fashioned baseboards, so that your eye will be less likely to pick them up at first glance. There are those who argue, however, that a radiator is a radiator, and any effort at concealment just ends up making them more noticeable. This school of thought would rather see one big honest radiator under a window than many feet of baseboard unit.

We were very excited about the radiators we found for our job. Not only did they leave the baseboards and other woodwork clear of obstruction, they were also delicate and beautiful objects in themselves. Like the electrical switches, they are not obtrusive, but when you do look at one, it's a pleasant surprise. We bought a whole variety of different configurations for different rooms, and we're happy with all of them.

The more traditional radiating devices in our units are, of course, the fireplaces and woodstoves. We were able to restore every original fireplace in the Main House, so that almost every room has one. Even if they aren't really efficient as space-heaters, they will certainly be cozy and romantic on cold winter nights. Each of the other units has either a new fireplace or a woodstove or

both. We installed the Woodshed Unit's stove in the big old fireplace of the original kitchen, where it can cook oatmeal for yet another generation, and in the Woodshed living room there is a new fireplace back-to-back with one in the Icehouse Unit. The Barn's fireplace is larger and faced in fossil-stone; behind it, in the Stable Unit, we installed a slender, cylindrical woodstove reminiscent of graceful European parlor stoves.

Choosing the little hardware like hinges, knobs and handles turned out to be fun. We didn't know exactly what the original hardware had been like because it had all been

stolen before we ever saw the place, but we knew what was typical of the period. There are hardware manufacturers who specialize in top-quality reproductions of old models, and we could choose from many possibilities. For door hinges and handles, we chose solid brass because of appearance and durability. In the Main House, these were replacement items, so we tried to come as close as possible to what the originals must have been like. Our final choices were small, elegant round brass knobs for the interior doors and a lovely brass thumb-latch pull for the outside of the front door. We felt that the porcelain knobs typical of Victorian interiors would not have been visible enough in this case, but we didn't want something too ornate. The front door handle is a faithful reproduction of an historical design, with a modern lock mechanism. This lock has a particularly durable spring and a safety release on the inside, in case of fire or emergency. We were so impressed by it that we felt all the front doors ought to have one.

The door hinges we settled on for the Main House were solid brass with an immovable center pin. This meant that the person installing them had to attach one side to the doors, then shim up the doors in the frames to exactly the right height, and finally screw in the hinge to the jamb—not

an easy operation with heavy old doors. A normal hinge will come apart into two pieces which can be attached separately, making it much easier to hang and unhang the door. The hinges we chose were so pretty, however, that we were willing to endure the added aggravation.

As you can see, thousands of decisions are necessary in order to get the details of your old house just the way you want them. These choices can be fun, especially if you're not fighting the clock every minute, and you should remember that they are very important in determining how the house will look when you're done. So don't get upset when it seems as if you have to spend weeks going through product catalogs—every decision you have to make represents an opportunity.

Chapter 10

Landscape

From the very beginning, we felt that one of the Bigelow House's greatest assets was its setting. The house was designed with a belvedere on the roof, floor-to-ceiling windows, an outside deck—all of which clearly express great expectations for the surrounding acres of meadows and trees. We knew that special consideration would have to be given to the landscaping. Simply putting any money left over from rehabbing into some minor site improvements would not be enough. Because a comprehensive design needed to be planned in which the building and the landscape were complimentary, we felt that we should hire a landscape architect.

Landscape architecture encompasses the design and layout of everything on a property that is not actually a building, although a landscape architect will sometimes advise how a building would be best situated to take advantage of its natural surroundings. Our landscape architect at the Bigelow House was brought in during the early stages of design in order to work closely with the architect and to be aware of the actual extent of the planned transformation.

A well-designed landscape makes use of sculptural massing, spaces, vistas, color accents and the effective combination of light and shade. It is also important that the plants that are chosen are effectively combined to

Color, texture, solids and voids, light, shadow, and even fragrance make up the composition of this small garden in New Jersey.

With completely different materials, this garden in Paris designed by Isamu Noguchi also makes a complex arrangement of color, texture, shape and light.

provide something interesting at any time of the year. This planning is especially important in a climate like New England's where the sight of an evergreen in a field of snow, or snowdrops in the bleak early spring can do so much for the cold weather blues.

Good design can make something as totally practical as a walkway or a driveway into something which actually enhances a property. A landscape architect may suggest, for instance, a curved path made of fieldstones, or a driveway paved in brick or pebbles. The average homeowner may just automatically choose the most common material— such as asphalt—because he is unaware of all the alternatives.

A good landscaping job, however, whatever its size, should be based on reason and form: these are the very bones of the design. Unusual materials and beautiful flowers are just the superficial aspects—it is the guiding design behind their use which counts in the end.

Beneath the apparent informality of the landscaping, a firm hand had been kept in the general design. This can be seen particularly in the control of the main spaces, which fall roughly into three categories: (1) the meadow; (2) the visitors' parking court; and (3) the courtyard. Each of these spaces is an

entity with its own scale, and is not allowed to compete with, or obtrude upon, the other spaces.

The Preliminaries

Our landscape architect for the Bigelow House was Tom Wirth of Sasaki Associates, Inc. of Watertown, Massachusetts—a firm nationally known for its high-quality landscape design. Tom inspected the site and came to much the same conclusions that we had in our research. Even before its years of decline, the house had been quite austere, without most of the customary details of the period. This led Tom to believe that the original landscape treatment would have been of a similarly spare character. Old photographs of the house had shown the wooden deck, flush with the grass, extending around most of the main house. This was to become an important element in the new landscape design. Unfortunately, the photographs did not show very much of the original landscape, and there was no other evidence of what it had looked like. But such clues as there were tended to support Tom's theory.

The design, then, proceeded on the assumption that the original landscaping was natural rather than formal. This character would be retained and enhanced in a layout tailored to the needs of the new owners. It would require a minimum of maintenance, and as many as possible of the original trees and shrubs would be kept. New plantings would be selected from species that were in common use around 1890, and the wooden decks which had originally served to link the house with the landscape would be extended to all the units.

It was Tom who first drew our attention to how carefully the entire building complex had been sited. ▶ Up to this point we hadn't given too much thought to how the entire estate had been put together, or why the house was shaped the way it was. It now began to seem that the placement of each part of the building was done for a very good reason. Apart from making the most of the sunlight, the southwest summer breezes, and the view, there were also many less obvious benefits: the stables were located to allow the wind to blow the animal odors away from the living and cooking areas, for example.

All this purposeful siting led Tom to believe that a landscape architect may have been involved in the original layout. It would most likely have been done through the office of Frederick Law Olmsted (the designer, among other things, of Central Park in New York and the Emerald Necklace

The Queen's Garden at Hampton Court in England is a formally planned garden of clipped hedges, geometrical paths and distinct beds of color and bloom.

"Natural" planting, as in this garden in Winchester, Massachusetts, imitates the informal juxtaposition of plants in nature, avoiding symmetry and obvious pattern. This casual appearance, however, is not casually planned or casually maintained.

system of parks in Boston) who had been a personal friend of both Richardson and Dr. Bigelow. No business records have been found to confirm this, but there was good reason to believe that the great landscape architect might have contributed some ideas casually in a conversation, or with a sketch done on the back of an envelope or a napkin, since that was the sort of thing he was known to have done on other occasions.

A Sense of Privacy

A problem that had been on our minds all through the design was how to provide as much privacy as possible for the people who were to share the building. Because the site was so idyllic and quiet, it became especially important that the owners of the units would not have to live in excessively close proximity, as that would create needless tensions. Jock had kept the entrance to the Barn Unit away from the courtyard in order to avoid congestion there. He had also sited all the living rooms so that more than half of their windows looked outwards towards the grounds rather than inwards to the court-yard, with the hope that this would draw attention away from it as an obvious center. This outward-oriented alternative was to be extended to the landscape by the use of more

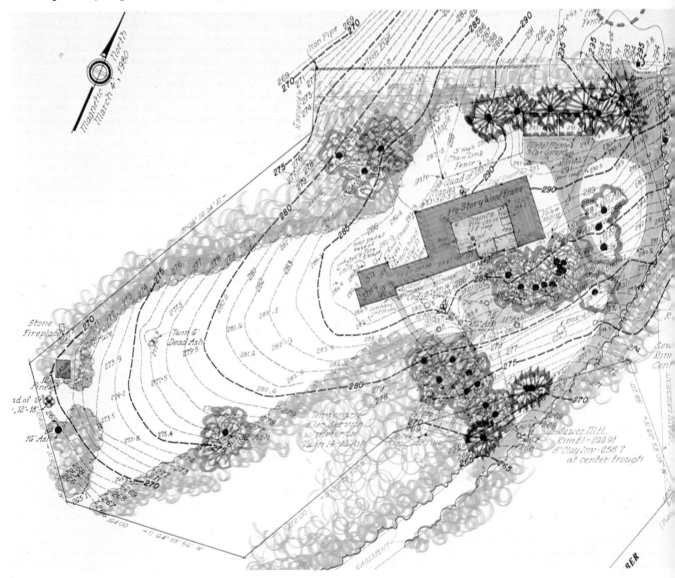

wooden decks, so that each of the units now had its own private outdoor living area, located to offer as much privacy as possible, and to be screened by planting. Also, each deck, as well as each front entrance, was connected to the new garage unit by one of a system of gracefully winding brick paths.

The Walkways

From the main spine of the entrance road, winding over the whole site, ran a system of pathways which connected the front and rear entrances of the units with the owners' garages and also provided some charming

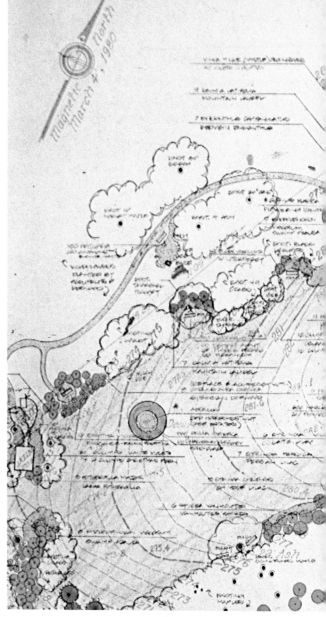

Spring flowering trees, Dutch bulbs, fall foliage, and evergreens will provide color at the Bigelow House all year. Rhododendron, dogwood, narcissus, and snowdrops (all shown above in other locations) are low maintenance perennials which bloom in the springtime. In summer, lilacs and other flowering bushes will succeed the spring

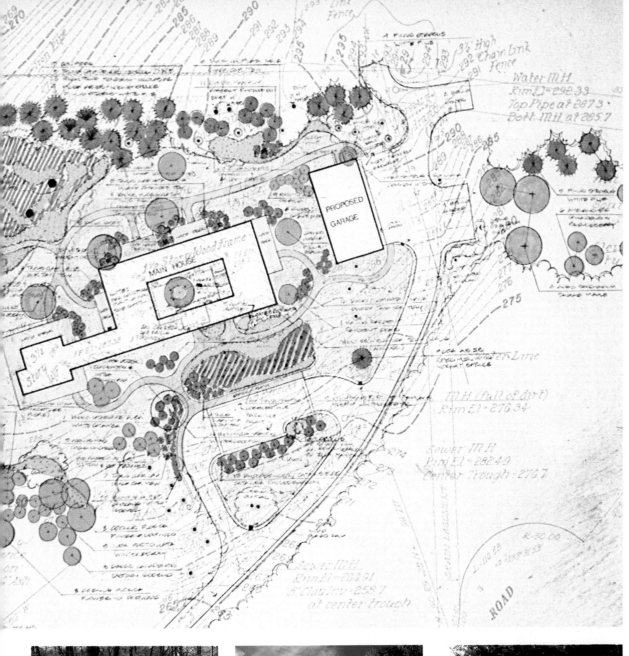

blooms, and in the fall, New England's famous leaf colors will be the main attraction. Winter color will depend on the shades and textures of evergreens, fall berries, and the birds which visit them.

walks through the woods. One path led from the visitors' parking lot to the units through some particularly attractive planting—a "greeting planting," as Tom put it, in which evergreens were predominant so that the effect would be cheerful in winter as well as in summer.

A dark red brick similar to that used in the parking lot was used to surface the paths—laid this time in a running bond instead of a pattern.▼ There was no evidence that any paved paths had been in the original layout, because the house had been intended mainly for summer use, and therefore expecting little traffic. But now, with increased inhabi-

tants and year-round use, some sort of durable surface was necessary. Gravel was rejected as being too unstable for a hillside, asphalt as not having been widely used in the 1890s, and stone pavers as too expensive. This left brick as the logical choice.

For further privacy, and total escape from the building, the brick paths also connected with a long extension which meandered southwards down through the trees to a small gazebo. At various points along this path there were niches, or sitting areas, planted with flowering deciduous trees and shrubs. The gazebo had been part of the original layout and was reached by walking

250

across the meadow; by approaching it now through the enclosure of the woods, the sudden open vista over the landscape was very dramatic. The gazebo was rehabilitated, linked with a new bluestone terrace and surrounded with new shrubs and flowers carefully chosen for their flowering succession. A small barbeque for picnics was also reconstructed.

The approach to the Main House had originally been a narrow cart track climbing along the hillside, with a sharp hairpin bend before the last stretch to the house. This still worked well, but needed a little widening at places along the road, and particularly on the corners. Halfway up the final stretch, convenient to the units, a place was made for visitors to park their cars by hollowing a platform out of the hillside between the house and the drive with plenty of room for a car to turn into it easily. To give the parking area a ceremonial, welcoming appearance, it was paved with patterned brickwork and separated from the asphalt of the driveway by a stonework border. It was also set slightly into the hillside in order to keep the cars more or less out of sight of the house. A low freestone retaining wall served a function similar to that of a *ha-ha*—the invisible sunken wall that formed a barrier around

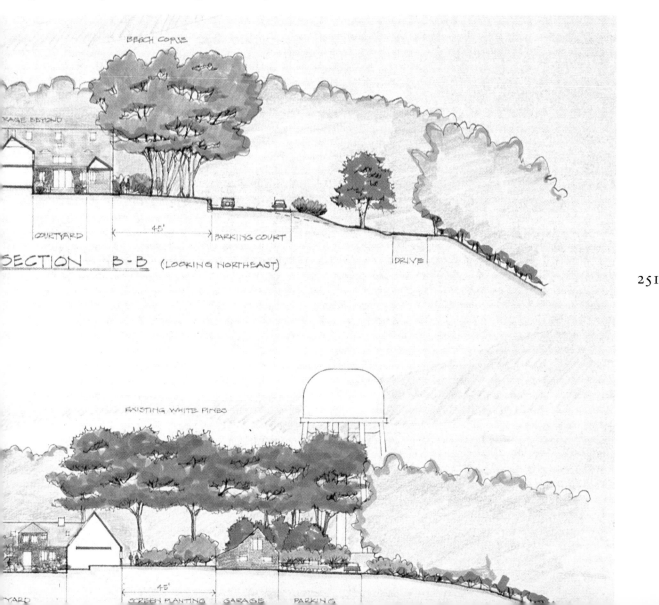

SECTION B-B (LOOKING NORTHEAST)

many eighteenth century English country houses. Uphill from this, an existing beech copse, reinforced with new planting, further screened the cars from the house. It also helped hide the roofs of neighboring houses, so that the views from this side of the building were unobstructed all the way to the horizon.

The Garage

A major problem in landscape planning is the location of a garage for the residents' cars. At the Bigelow House, a long shed building at the top of the hill in the northwest corner of the site had been used for this purpose in recent years. Earlier, it had housed carts and agricultural implements; we found an old hayrake there with wooden wheels—a sure sign that the meadow had actually been cut for hay at one time. Because of its dilapidated condition, however, and since it was not really part of the original architecture, ▼ we decided to demolish the shed and build a new garage facility somewhere in that area—the least visible part of the site. Earlier in the design stage we had discussed the possibility of using the south slope of the Main House as a solar collector. The large, steeply-pitched roof (approximating the New England ideal of fifty-four degrees) would have made a most

efficient collector; also, many members of the Historical Commission felt that this was just what Richardson and Dr. Bigelow would have done had they been alive today. This idea was dropped, however, because of the radical change it would have made in the appearance of the house on its most visible side. The south-facing roof over the Barn Unit did not have enough area, and it had the important eyebrow windows in it.

When the new garage was proposed, it provided the perfect opportunity to take advantage of the sun without spoiling the roof lines of the main building. The new unit became the garage/storage/solar collector building, and was sited at right angles to the old shed in order to have a south-facing roof. This meant that it would run parallel to the Barn Unit and just to the north of it. ▶ It was kept as low as possible and partially screened by Dwarf Japanese Yew.

Plant Materials

Our guiding principle in the selection of plant material was to use species that were known in the 1890s—however we did not want to, by any means, suggest any kind of restoration or recreation of a period garden. Indeed, it was very doubtful that this very practical complex of buildings—in some ways more resembling an added-on New

252

England farm than a Victorian summer home—would have ever had such frivolities as knotgardens or ornamental flower beds! Also, as Tom pointed out, by the end of the nineteenth century the fashion of growing things that were only ornamental had been largely replaced by a scientific interest in plants which included their careful naming and derivations. Actually, it was to the Landscape Garden of the eighteenth century that we turned for guidance.

Of course, we were not planning to reproduce the grounds of an English country house—we merely borrowed some of its design principles. The Landscape Garden was an English reaction to the highly formal French garden, such as found at Versailles. This was usually laid out symmetrically around a grand axis, from which it was meant to be seen. Everything was controlled by a rigid geometry, even the shrubs, which were clipped into sculptural shapes. The English approach, in contrast, went back to nature for its inspiration. The Landscape Garden looked natural, but was, in fact, contrived. Each informal group of trees was actually very carefully studied and each casual color accent very carefully placed.

This is what we tried to do on a small scale at the Bigelow House. The principal

tree groupings were already well established, and so we simply modified them for our different needs. The vista to the south over the meadow, for instance, was slightly enlarged by trimming back some of the newer trees at the side boundaries, and by removing others at the lower end down by the gazebo. ▼ Most of the trees around the house were kept to provide useful shade in the summer, and to screen the roofs of neighboring houses. Several massive European beeches were "dug out of the woods" by removing nearly a half-acre of dense undergrowth. The only major tree planting we did was on the western boundary of the site, near the house, and it consisted of nineteen evergreens—mainly hemlock, spruce, fir and pine—forming a screen between the house and the only other house that was at all close by.

In planting shrubs and flowering plants, some planning and strategy is necessary to get the maximum effect for the longest period. Tom concentrated most of his efforts along the sides of the space which would be seen by most people—the meadow. The trimmed-back edge of the vista was planted with low shrubs and flowers which would be seen to maximum effect against the darker trees and serve as a frame to the view, with occasional color accents. These plants were

carefully chosen to thrive in either shade or sun, depending on their location, and also to provide attractive color combinations at all times during the growing season. Here is an annual diary of events on the edges of the meadow:

Late Winter
Witch Hazel

Early Spring
Cornelian Cherry, Daffodils, Forsythia, Shadblow

Spring
Azalea, Flowering Cherry, Dogwood, Enkianthus, Lilac, Rhododendron

Early Summer
Chinese Dogwood, Spirea, Yellowwood

Summer
Summer-Sweet, Swamp Azalea

Other shrubs and plants were located either singly or in combinations at selected places throughout the site, such as the niches at the side of the path to the gazebo. Small *incidents* such as the Smoke Bush by the path to the main house were strategically placed to enliven the environment—sometimes at quite unexpected places. Another such example, very much in the natural tradition, was a boulder which proved too difficult and

254

expensive to remove, and now gives an interesting accent by simply being there.

The most detailed planting was reserved for the one place in the whole project where it would receive the most attention—the courtyard. ▶ Our aim was to create a space that would be an attractive entrance both to the owners and their visitors; a shady, quiet place to sit in the warm seasons; and yet allow for privacy in the adjoining living units. The same paving brick that was used in the paths and the parking lot was continued into this area with the perimeter being given over to planting, granite steps and stoops to the entrance doors of the living units, and the

wide wooden ramp at the greenhouse of the Barn Unit. A little off-center in the paved area was a ten-foot island planted with ground cover, and featuring a medium-sized shade tree in the middle. This was an exotic species related to the Locust, and it gave enough shade from the summer sun without darkening the interiors of the unit too much. The brick-paved area was framed by a granite edge, echoing the visitors' car park.

The planting ranged mainly along the north and west walls of the courtyard in order to make the most of the sun. As would be expected, it was also carefully chosen for its flowering succession. It included some of the same plants that were used in the meadow plantings but also a number of perennials and bulbs.

Access to the courtyard via the brick pathways was from the east—through the fourth side of the enclosure. Half of this wing had originally been a storage space, but it was now opened up to form a cloister. The openings which had framed the sliding doors to the storage space and the old carriage entrance to the courtyard were now opened up to form the new entrance. A bench was located in the cloister area between the two openings, giving a picturesque framed view of the courtyard from the shade.

256

Kerria japonica *is a flowering shrub typical of those popular at the end of the 19th century. Its yellow blossoms decorate long, drooping branches like those of the forsythia bush.*

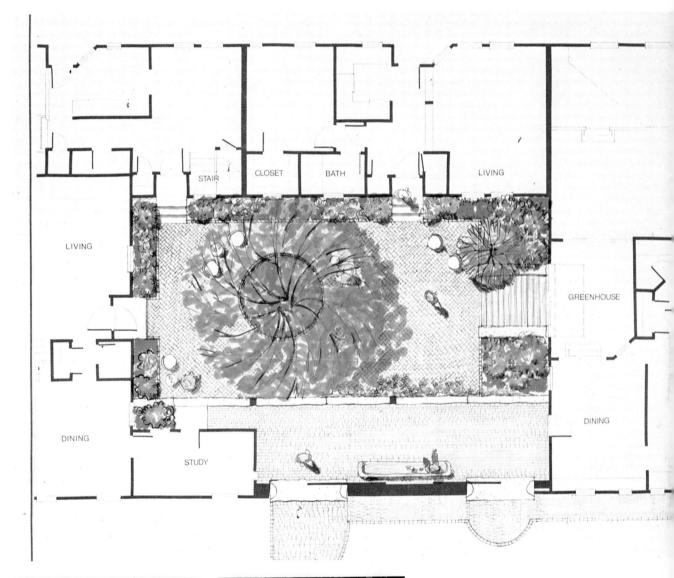

STAIR CLOSET BATH LIVING

LIVING

GREENHOUSE

DINING

DINING

STUDY

Pyracantha, or fire thorn, produces such exuberantly orange berries that it is a valuable color accent in any fall garden.

Lilac, one of the old-fashioned favorites we planted at the Bigelow House, comes in a huge array of popular variations. Most are extremely fragrant and bloom luxuriantly in early summer. This is a Chinese tree lilac, which blooms a little later.

Chapter 11

Five New Houses

When Dr. Bigelow commissioned the house for his Oak Hill property in 1886, one of the things he particularly liked about the spot was what he called its "audible silence."

No one could have listened to any silence here for the last year and a half. Chain saws, cement mixers, sledge-hammers, power tools of every description, bullhorns: you name it, we made noise with it.

Now, however, the silence has returned. The red-shingled structure at the top of the hill is now five new homes, ready and waiting for their new owners.

The approach to the house is a newly paved drive, which turns more sharply than

before. A fine new fence along the right-hand side of the drive adds to the feeling of privacy.

If you're a visitor, you can turn left at the brow of the hill and park in a brick-paved parking court. A flight of steps leads from there to the house.

A Bigelow House resident would drive right up and around the north end of the barn, to the parking garage that serves all five units. This is the one totally new building we put up and it is finished in red-stained cedar shingles like everything else. There is space for five cars and numerous bikes, baby carriages, lawn furniture—even a workshop or two. All the utilities (gas, electricity, telephone, drainage) are located underneath the garage. We buried all the wires so as not to mess up the countryside, and we centralized all the cables and pipes here.

North of the garage building there is a graceful turnaround, or extra parking area, with a grouping of red oaks in the middle. We left as many trees as possible.

Walking around the garage, on the west lawn, there are two cedar decks along the side of the building, which belong to the Stable and the Icehouse. Beyond the lawn, a stone-dust trail winds through the woods that fringe the western side of the property,

like H. H. Richardson's drawing of a century ago. It is again clad in cedar shingles from top to bottom, and two coats of protective stain have given it a warm russet-red hue. The narrow trim around windows and doors is painted dark green. The sunbath has a copper-hipped roof now, replacing the awkward shed-type roof we began with. Another note of copper is the new hat on the turret, with a bulbous finial at the top just as the architect originally sketched it.

Up on top of the roof you can see the belvedere with its reconstructed railings and canopy supports, and the old brick chimney, newly pointed.

Cedar decks surround the Main House. In a way, these do alter the public facade of the house, which we have worked hard to keep consistent with the original plan. But no one actually knows whether or not Richardson envisioned decks. The old Storrow family photographs did show decks, and we thought they were a good addition. After all, Dr. Bigelow was a great man for fresh air. He had some say in the building of a belvedere on his roof and the floor-to-ceiling windows in his living and dining rooms. It seemed only logical to step out through those windows onto a deck.

and it opens out into a number of clearings.

The trail brings you to the far southwest corner of the south lawn, where the gazebo stands. From here you can really appreciate the effects of Tom Wirth's landscaping design. Flowering bulbs are everywhere, thousands of them. Plantings of lilacs and laurels and rhododendrons ring the lawn, which is green and smooth with its acre of bluegrass sod. What a contrast from construction days, when it was an expanse of rutted mud dotted with vans and trailers, cables and camera booms.

The lawn sweeps up to the south face of the Main House—which once again looks

The Main House

Going around to the east side of the Main House, you come to the front door under the second-story projection with the carved lookouts. Turn the new brass doorknob and you're in a parquet-floored hallway that looks much the same as it did in Dr. Bigelow's day. The stairs go up ahead of you, their balusters carved in those three distinctive styles. Turn left into the living room and you find white painted walls and a gleaming hardwood floor. Here, and all through the house, the transition from the abandoned and disfigured shell we had found was remarkable. The old pine paneling has

been cleaned and replaced around the fireplace, and the fireplace surround itself is a clever combination of the old ceramic tiles and new ones that match beautifully. There's a bar in Dr. Bigelow's old cubby just to the left of the fireplace. And at both ends of the room are the tall triple-hung windows, rebuilt and in good working order. The room is sunny and fragrant with cedar of the paneling in the sunbath, which we've renamed the conservatory. It has ceramic tile flooring and we hope it is soon filled with flowers and plants.

In the Main House we were tremendously fortunate to have many original details worth

restoring. We were able to recreate others, and to replace still others with components of equivalent quality. The level of quality that prevailed in the Main House set the tone for the rest of the units.

Going back across the entrance hall, you come to the room that was most dramatically changed: the dining room/kitchen, formerly just the dining room. The two spaces are separated by a waist-high counter, but you can see all the way through from the old tiles around the dining-room fireplace to the solid oak cabinets in the kitchen. The kitchen has a ceramic tile floor, Formica countertops, and every modern cooking gadget. Yet the

cabinets maintain a sense of unity between the old and the new, being so similar to the woodwork throughout the house.

As you walk up the stairs to the second floor, you first see a guest room, the room above the dining room that was originally built for Dr. Bigelow's son. It has a bath and an enormous walk-in closet. Windows look out to the east and to the west. Back across the hall is the master bedroom, now expanded to fill the whole south side of this floor, including what was once Dr. Bigelow's bedroom and his study. There are two back-to-back fireplaces, and there is a stunning view across the south lawn. And here is

the turret, all its windows newly glazed, a fine setting for the new owners' whimsy.

The master bathroom, now housed in the projecting space above the front door, has a luxurious whirlpool bath and octagon-and-dot tiles, and a southern view from the angled-in window.

Up another floor, you find two smaller bedrooms: one faces west and one (with a fireplace) east. Their bathroom is tucked under the eaves across the hall. Through it, you reach a large storage space under the gable roof of the former servants' wing.

From the third-floor hall a small spiral staircase takes you up to the attic, virtually unchanged since the house was built except for extensive insulation. From there, a final set of steps winds upwards to the belvedere hatch.

Altogether this is a cozy house. It is roomy but not overwhelming, nicely planned, endowed with ample storage space, and graced with a feeling of privacy and solitude. It's a good blend of restoration and adaptation. And it manages to keep the spirit of the house as it originally existed without sacrificing any contemporary convenience.

The Woodshed
The largest of the three units around the courtyard is the Woodshed. It has doors

both into the courtyard and out onto a deck that faces south. We reach the deck by walking out through the tall window at the west end of the Main House living room, turning right, and following the house around the corner.

The living room of the Woodshed Unit is bright with sunlight and rises two stories. One wall holds a large brick fireplace with an energy-efficient heat-exchanger device, and a storage niche for logs. Overlooking the living room like a balcony is the semi-circular landing of the Woodshed's dramatic curved stairway. Leaving the living room, you move into the area that was once the kitchen, servants' hall, and service area of the Main House. There is now a large kitchen with a fine east view, a fireplace, ample cabinets, and a large center work table.

Next comes the dining room, also facing east. Off the dining room is a small, private room—a perfect study or playroom—with a tiny peephole window that looks into the courtyard entrance under the portico.

If you return through the dining room to the foot of the stairs and go up, you find two large bedrooms, one with a fireplace. The northernmost room looks east (over the hills) and north (into the courtyard). On this floor, there are two full bathrooms, several closets, and a laundry area.

The Courtyard

We leave the Woodshed Unit by its door on the north side, into the courtyard. This has now been paved with bricks, the same type used with such success in Boston City Hall's plaza and the redevelopment of Boston's Faneuil Hall/Quincy Market. The whole rectangle is edged with shrubbery, ground cover, and flowering bulbs—and these beds are set off by a border of granite paving stones. In the center is another growing area, with a flowering dogwood tree and some smaller plantings. The overall effect is one of simplicity and peacefulness, and it will be easy to maintain.

The Icehouse and Stable Units, on the west side of the courtyard, were extensively rebuilt inside and out, and have new doors, both into the courtyard and out to the western lawns. There, each unit has its own deck. Beside the deck doors are windows angled to catch the southern sun and view, like the one window in the second floor bath of the Main House.

The Stable

The Stable is the smallest of the five units. We wanted to be able to offer one unit for a relatively moderate price. Even so, here, as everywhere, we were fortunate to have been given superior materials. Careful attention to detail and solid construction lend a feeling of comfort and quality to the place. There are two big rooms here, as well as a kitchen and bathroom. It is as compact as a suite on a ship. Yet the living room has exposed beams, a brick wall, and a Petit Godin enameled wood stove.

The Icehouse

The Icehouse is quite a bit larger. It extends all the way to the southwest corner of the rectangle that bounds the courtyard. Its living room is down a couple of steps from the rest of the ground floor, and boasts a wall of old bricks, a large fireplace (with a heat exchanger like the one in the Woodshed), and a

269

great southern view through three large windows. The dining room looks west, onto the deck.

Going up the stairs—which are next to the courtyard door of the unit—you find yourself on the long second story under the line of the gable roof. Starting at the southern end, we find a bedroom with big square glass doors to the south that open out to a small balcony. There are two skylights, a laundry and bathroom area, a study with three skylights (where the stairs come up), and a bath and dressing area with huge closets. Finally, at the north end of the Icehouse second floor is its master bedroom,

with views of Blue Hill in the distance and the courtyard below.

The Barn

The Barn is the place where we could really play with space and try out our most innovative ideas.

Here, the door which used to admit horses and farm equipment is now all glass, but a ramp, like the one the horses must have used, remains. We first enter the greenhouse, and it—like the Main House conservatory—is fragrant with cedar paneling. Sunny, with a full southern exposure and sheltered location, it should keep itself pretty warm. It has an old soapstone sink for potting and planting, a relic we salvaged from another rehab project.

Going up a flight of steps to our right, we come to the dining room and then, turning left again, the kitchen. This kitchen is definitely designed with the serious cook in mind. There's a ceramic tile floor, a center work area, an enormous expanse of Corian countertop, wood cabinets, a four-burner range and microwave and conventional ovens. Both kitchen and dining room, modern equipment notwithstanding, have the small square windows the horses used to look out of. This was a barn, after all.

From the kitchen you can walk into the

front hall of the Barn, passing a utility center which houses, among other things, a heat pump. The Bigelow House incorporates just about every practical method of heating and cooling.

The Barn has a study and a sunken living room, separated by a low partition, and illuminated by light that comes down through a two-story shaft. In the living room, one full wall is bookshelves and cabinets, with a dramatic rough-hewn fireplace made of fossilstone. At the west end of the room, tall windows look out across the lawn to the woods. Also at that end, you can go up a spiral stair to the second floor. There, too,

the view is spectacular, through a window that goes right up the middle of the west wall (where a hay-door used to open into the loft) to the point of the roof. This window illuminates both the guest bedroom and a small study. The study also has a skylight; the guest bedroom has two. In the guest bathroom is one of the famous eyebrow windows we admired on our first visit to Oak Hill.

We move from the guest bedroom back east toward the upstairs hall. The front stairs of the Barn Unit lead up to this hall in a most inviting but compact fashion. The master bedroom, with its ample closets, now occu-

pies the east end of what began as one vast hayloft. From the bedroom you can still look down the entire length of the barn. Just past the adjoining master bath is the hall with the laundry alcove on the left facing the main stairwell. The second bedroom, its bath, and the study are beyond. The rooms up here have a wonderful feel of airy space, thanks to the fifteen-foot ceiling punctuated by skylights in each of the bays. And in the master bedroom, where the hay door used to be there is now a huge recessed sliding glass window with a view to the east through the tree tops.

We have worked quite a transformation here, and, in this book as well as on television, we have tried to explain both the grand plans and the detailed decisions as we went along. It has been exciting to see the rejuvenation of the Main House, and the innovation in the other buildings. As the grateful recipients of other people's generosity, we were able to use high-quality materials quite inappropriate to most people's budgets.

But we have preserved an important piece of the past, and given it new life, and found a satisfaction we hope you've been able to share.

274

Business and Legalities

Cassiday, Bruce. *The Complete Condominium Guide.* New York: Dodd, Mead and Co., 1979.

Heatter, Justin W. *Buying a Condominium.* Cambridge, Mass.: Financial Strategies, Inc., 1980.

Home Buyer's Vocabulary. Washington, D.C.: U.S. Department of Housing and Urban Development (pamphlet), 1979.

Mencher, Melvin, editor. *The Fannie Mae Guide to Buying Financing and Selling Your Home.* Garden City, N.Y.: Doubleday and Co., Inc., 1973.

Preservation and Building Codes. Washington, D.C.: National Trust for Historic Preservation, 1975.

Questions About Condominiums: What to ask before you buy. Washington, D.C.: U.S. Department of Housing and Urban Development (pamphlet), 1980.

Reilly, John W. *The Language of Real Estate.* Chicago: Real Estate Education Co., 1977.

Vitale, Edmund. *Building Regulations.* New York: Scribner's Publishing Co., 1979.

Wise Home Buying. Washington, D.C.: U.S. Department of Housing and Urban Development (pamphlet), 1980.

H. H. Richardson

Hitchcock, Henry-Russell. *Architecture of H. H. Richardson and His Times.* Cambridge, Mass.: The M.I.T. Press, 1966.

Mumford, Lewis. *South in Architecture.* New York: Da Capo Press, Inc., 1967.

O'Gorman, James F. *H. H. Richardson and His Office: Selected Drawings.* Cambridge, Mass.: The M.I.T. Press, 1979.

Scully, Vincent J. *The Shingle Style and The Stick Style*, revised edition. New Haven: Yale University Press, 1971.

Van Rensselaer, Marina G. *Henry Hobson Richardson and His Works.* New York: Dover Publications, Inc., 1979.

General Books

Carrington, Merrill Ware. *Design Guidelines: An Annotated Bibliography.* Washington, D.C.: National Endowment for the Arts, 1977.

Jandl, H. Ward, and Marsha Glenn. *A Selected Bibliography on Adaptive Re-use of Historic Buildings.* Washington, D.C.: National Park Service, Technical Preservation Services, 1976.

Kettler, Ellen D., and Bernard D. Reams, Jr. *Historic Preservation Law: An Annotated Bibliography.* Washington, D.C.: National Trust for Historic Preservation, 1976.

Periodicals

American Preservation: The Magazine for Historic and Neighborhood Preservation. Bi-monthly. Bracy House, 620 East Sixth, Little Rock, Ark. 72202.

Bulletin. Quarterly. Association of Preservation Technology, Box 2437, Station D, Ottawa, Ontario, Canada K1P 5W6

Historic Preservation. Quarterly. National Trust for Historic Preservation, 1785 Massachusetts Ave., N.W., Washington, D.C. 20036.

The Old-House Journal. Monthly. The Old-House Journal, 69A Seventh Ave., Brooklyn, N.Y. 11217.

Technology and Conservation. Quarterly. The Technology Organization, Inc., 1 Emerson Place, Boston, Mass. 02114.

Technology

Browne, Dan. *Alternative Home Heating*. New York: Holt, Rinehart and Winston, 1980.

McGuiness, William J., Benjamin Stein and John S. Reynolds. *Mechanical and Electrical Equipment for Buildings*. New York: McGraw-Hill, 1980.

Moore, Harry B. *Wood-Inhabiting Insects in Houses*. Washington, D.C.: Department of Housing and Urban Development and Department of Agriculture, 1979.

Shurcliff, William A. *Solar Heated Buildings of North America: 120 Outstanding Examples*. Harrisville, N.H.: Brick House Pub. Co., 1980.

Renovation and Rehabilitation

Bullock, Orin M. *The Restoration Manual*. Norwalk, Conn.: Silvermine Publishing Co., 1966.

Bunnell, Gene. *Built to Last: A Handbook on Recycling Old Buildings*. Washington, D.C.: Preservation Press, 1977.

Cantacuzino, Sherban. *New Uses for Old Buildings*. New York: Watson-Guptill Pub., 1975.

Diamonstein, Barbaralee. *Buildings Reborn: New Uses, Old Places*. New York: Harper and Row, 1978.

Ferro, Maximillian L., *How to Love and Care for Your Old Building in New Bedford*. New Bedford, Mass.: Office of Historic Preservation, 1977.

Jacopetti, Ronald and Benjamin Vanmeter. *Rescued Buildings*. Santa Barbara, Calif.: Capro Press, 1977.

Nielson, Sally E., editor. *Insulating the Old House: A Handbook for the Owner*. Portland, Maine: Greater Portland Landmarks, Inc., 1977.

Reiner, Laurence E. *How to Recycle Buildings*. New York: McGraw-Hill Pub. Co., 1979.

Stephen, George. *Remodeling Old Houses, Without Destroying their Character*. New York: Alfred A. Knopf, Inc., 1972.

Thompson, Elizabeth Kendall, editor. *Recycled Buildings, Renovations and Re-uses*. New York: McGraw Pub. Co., 1977.

Wrenn, Tony P. and Elizabeth D. Mulley. *America's Forgotten Architecture*. New York: Pantheon Books, 1976.

Index

278

281

Chapter 1

Page 2 (l) photo: Steve Rosenthal; (m) photo: courtesy devel/bldr: Mac Grant; (r) photo: Tad Goodale, courtesy Children's Museum
4 (l) arch: Gelardin/Bruner/Cott, Inc., photo: Greg Heins; (r) arch/devel: Caivano & Associates, photo: Bill McDowell
5 (l) photo: courtesy Historic House Association of America & Dr. S. Mike Mahan; (r) photo: Betsy Fuchs
8 photos: from Storrow family album, 1896
10 (l) photo: archives of the Society for the Preservation of New England Antiquities; (m) photo: archives of SPNEA; (r) photo: Jack Boucher for National Architectural & Engineering Record
11 photo: Walter Smalling, Jr., for Heritage Conservation & Recreation Service
12 (r) photo: courtesy Henry H. Richardson III
17 (l) devel: Pittsburgh History & Landmarks Foundation, int. design: Robert Sherman & Associates, photo: Clemens Kalischer © 1980; (r) arch: Witsell & Evans, photo: Greg Hursley
18 (l) arch: Booth, Nagle & Hartray, Ltd., photo: Barbara Crane; (r) arch: Joseph Power, owner/devel: Michael L. Ainslie, photo: Joseph Power
20 photo: courtesy Lunar and Planetary Institute

Chapter 2

22 arch: Backen, Arrigoni & Ross, project designer: Miles Berger, photo & illustr: Walter Thomason & Associates
23 (l) photo: Walter Smalling, Jr., for HCRS; (r) photo: courtesy Baltimore Dept. of Housing & Community Development
24 (l) devel: Calumet Industrial District Company, photo: Hunt McKinnon; (r) devel/owner: The Pomerleau Agency, photo: Carolyn Bates
25 arch: Stickney & Murphy, photo: courtesy Seattle Fire Dept.
26 (l) development team: Whitefield Associates; (m) arch: William H. Short, Short and Ford, photo: Otto Baitz; (r) photo: Melvyn E. Schieltz
27 (l) arch: Short and Ford, photo: Otto Baitz; (r) bldr/devel: Waterstreet Associates, photo: Walter Smalling, Jr., for HCRS
28 (l) arch: Richard Boast, contractor: Wexler Construction Co., Inc., photo: Lilian Kemp; (r) devel: Artspace, Inc., Cora Able & Thomas Fodor, photo: courtesy Cambridge Historic Commission
29 (l) arch/devel: DuBose Associates, Inc. & The Hartford Architecture Conservancy, photo: Roger Dollarhide; (r) arch: William Kessler and Associates Inc., photo: Balthazar Korab
30 arch: Jay Leavitt & Associates, devel: The City Living Group, Inc., photos: Dick Klein
31 (l) photo: Roger Whitacre; (r) arch: The Taft Architects, photo: courtesy Galveston Historical Foundation, Inc.
32 (m&r) arch: Fletcher Dean and Associates, photos: Greg Hursley
33 arch: Crissman & Solomon, (r) photo: Steve Rosenthal
34 arch: Nagle, Hartray & Associates Ltd., photos: Tom Van Eynde
35 arch: Hefley Stevens Architects Inc., (r) photo: Jerry Morganroth
37 arch: Mary Otis Stevens, devel/contr: Sydney & Noyes Associates, photo: Mark Landsberg
38 (l&m) arch: Bell, Klein & Hoffman, photos: courtesy Franklin Savings Association; (r) arch: Hammer Kiefer and Todd, Inc.
39 arch: Cambridge Seven Associates, Inc., (l) photo: courtesy San Antonio Museum of Art; (r) photo: Steve Brosnahan

Chapter 3

40 (l) photo: George Stephen; (r) photo: Melvyn E. Schieltz
42 (l) arch: Kenneth Schroeder Associates, photo: Sadin/Karant; (r) restoration: Clifford Zink
43 arch: Zwibald/Clarke, photo:

283

Chapter 9

218 arch: Duncan McGowan, photos: Robert M. Aude
220 photo: Paul Dietrich
221 (l) arch: CBT/Childs Bertman Tseckares & Casendino Inc., photo: Robert Perron; (r) arch: Short and Ford, photo: Otto Baitz
224 arch: Myron Goldfinger, photo: Norman McGrath
225 (l) arch: Kenneth Schroeder Associates, photo: Ron Gordon; (r) photo: Allen Freeman
231 (l) arch: Unihab, Cambridge, Mass.; (r) restoration: Clifford Zink
232 (l) photo: courtesy Peter Chermayeff; (r) restoration: Clifford Zink, photo: Philip Smith
235 Alpes-Inox appliances from Kamenow, Inc., 823 Madison Ave., New York, photos from *Attention To Detail*, by Herbert Wise, Copyright © Quick Fox, 1979
238 (l) photo: courtesy Unihab, Cambridge, Mass.
240 (l) photo: Robert Perron

Chapter 10

244 photos: Gary Mottau
246 photos: (l) Gary Mottau; (r) Franklin Photo Agency, Nelson Groffman
248 photo: Franklin Photo Agency, Nelson Groffman
249 photos: (l&m) Gary Mottau;

(r) Tom Wirth
256 photo: Gary Mottau
257 photos: Gary Mottau

(l): left
(m): middle
(r): right
arch: architect
devel: developer
bldr: builder
contr: contractor

The main body of the book was set on the Linoterm in Bembo, a copy of a roman type cut by Francesco Griffo for the Venetian printer Aldus Manutius. It was first used in Cardinal Bembo's *De Aetna* in 1495. The type was composed by The Stinehour Press, Lunenburg, Vermont. The book was printed and bound by R. R. Donnelley & Sons Company, Willard, Ohio. Color separations were made by Tru Color, West Hollywood, Florida.

This book was designed by Lorraine Ferguson, Douglass Scott, and Christopher Pullman.

Supplementary photographs were researched by Eleanor Dietrich.